Adam and Evolution

Adam and Evolution

Michael Pitman

Introduced by Bernard Stonehouse

Rider

London Melbourne Sydney Auckland Johannesburg

Rider & Company
An imprint of the Hutchinson Publishing Group

17–21 Conway Street, London W I P 6 J D

Hutchinson Publishing Group (Australia) Pty Ltd
16–22 Church Street, Hawthorn, Melbourne, Victoria 3122

Hutchinson Group (N Z) Ltd
32–34 View Road, P O Box 40–086, Glenfield, Auckland 10

Hutchinson Group (S A) Pty Ltd
P O Box 337, Bergvlei 2012, South Africa

First published 1984
Text © Michael Pitman 1984
Introduction © Bernard Stonehouse 1984
Illustrations © Andrew English 1984

Set in Compugraphic Plantin by Colset Pte Limited

Printed and bound in Great Britain by Anchor Brendon Limited, Tiptree,
Essex

British Library Cataloguing in Publication Data
  Pitman, Michael
    Adam and evolution
    1. Life—Origin
    I. Title
    577 QH325

  ISBN 0 09 155390 3 cased
        0 09 155391 1 paperback

For Suzanne,
    Marianne
    and Emmanuel

Contents

Illustrations

Introduction

Michael Pitman, the author of this book, teaches biology in Cambridge. For a biology teacher his background is unusual; he read classics at Oxford, and after several years of self-education has recently taken courses in biology at the Open University. Unusual, but not unique; he follows an honourable tradition. J.B.S. Haldane, physiologist, evolutionist and geneticist of the last generation, had no formal training in biology either, but taught it memorably in lectures at University College, London, in seminars in the pub in Gower Street, and possibly to his widest and most appreciative audience through a regular column in the *Daily Worker*. Neither Charles Darwin nor Thomas Huxley, who figure in this book, were trained in biology; nor were Gregor Mendel or Goethe, or, more recently, Jacob Bronowski or Arthur Koestler. This did not stop them contributing substantially to biology as innovators, teachers or both.

His unorthodox background must have helped Michael Pitman to write this book; it might not have occurred to one formed in a more conventional mould. His thesis is that Darwinism, the biological concept of evolution by natural selection, is largely unproven and may be misleading. Why does this matter to anyone but biologists? Quite simply because Darwinism, though invented to explain phenomena in natural history, spilled over from the mid-Victorian crucible in which it was brewed and ran like wildfire through every other branch of philosophy. It penetrated even religion and morality to become, as clerical opponents feared it would, an aspect of common sense.

Today, we are all evolutionists, to a degree that might have been quite incomprehensible to our early nineteenth-century forebears. Anyone who believes that events take their course, that situations arise from each other without the constant, active interference of supernatural powers, owes something to the battle that evolutionists have fought and won. There were evolutionists long before Charles Darwin; his postulated mechanism of *natural* (as distinct from human or divine) selection gave evolution its

9

mainspring, and contemporaries the opportunity they needed to bring the whole concept of evolution into the mainstream of philosophy. If Darwinism turns out to be nonsense, the basis for much of our everyday thinking is suspect, and we may all have to look for something better.

Michael Pitman tackles Darwinism, and the twentieth-century 'neo-Darwinism' that arose from it, on their home ground of biology. Few biologists will disagree with his proposition that, in accepting Darwinism, we take a great deal on trust. Indeed I would suspect the wits of any biologist who, on serious reflection, accepted every aspect of Darwinism uncritically. But in *Adam and Evolution* the author points out just how little serious reflection anyone, biologist or not, gives to Darwinism these days, and just how much we take on trust when we take Darwinism aboard. He does it with wit and a nose for absurdity, stirring the dust with a welcome breath of fresh air.

When I took a degree in zoology a generation ago, nobody so far as I can remember *taught* us the Darwinian approach to evolution. The assumption that we knew and accepted it was implied in every statement put forward by every lecturer. Nobody tried to suppress anti-Darwinian views – there were simply no anti-Darwinian views of any consequence to be suppressed. Someone may have mentioned in passing the absurdity of Lamarckism. We knew of Fundamentalists, Creationists and others in the world who did not believe in evolution, far less in evolution by natural selection; there were Flat-earthers, too, and folk who believed that the pyramids held all the world's wisdom, but we didn't take them seriously either. No disciplined, cohesive counter-Darwinian viewpoint came across the lecture benches. Darwinism was assumed, much as chemists assume the existence of atoms, and theologians the existence of God.

I read Darwin's *Voyage of the Beagle* with delight, and as much of *The Origin of Species* as I had time for. Like everyone else in my group I decided that Darwin rules, and got on with more pressing matters. I do not complain that we were taught this way. Biology was changing rapidly; we had much to learn in three or four years of professional training, and many things, including Darwinism, had to be taken for granted. So as students we imbibed it almost without question, certainly without serious misgivings, like organic lead from the atmosphere or alcohol from gin and tonic.

A great deal of what I learnt as a student has since proved to be wrong, or at best relatively trivial; accepted wisdom has changed and is still changing fast. But the essential philosophy of self-determinate evolution flourishes as strongly as ever. Darwin still rules, and on the whole rules benevolently. Natural selection is still the working model that almost every professional biologist keeps at his elbow. It is still taught widely in schools, and still part of the wider common sense that everybody accepts and uses in daily life.

Perhaps, like organic lead, Darwinism has become a cumulative poison, though that would not be my view. It may well be a stimulant, but only those who fear it are likely to brand Darwinism a poison. An addiction, perhaps? Addictive it can certainly become. My own research and teaching have always been planned and interpreted in Darwinian terms; so have those of every other working biologist I know. We assume that animals and all other living creatures adapt to environments, that natural selection snaps at their heels to speed the process, that today's organisms are adapted through natural selection from different kinds of organisms that lived long ago, and that the fossil record, properly interpreted, can show us how it happened.

This is all a useful framework for the mass of knowledge that modern biologists must acquire but, as Michael Pitman points out, there is a hidden danger. Like almost every other working biologist I rejoice (well, nod with mild approval) when one of my colleagues comes up with evidence, honestly acquired, that fits the Darwinian model. Like others I suspend judgement on evidence that seems not to fit, and feel better when, with further thought, I find it consistent after all. Like others I tend to test the evidence against the model, and never really find time to test the model itself.

Honest workers do not bend the evidence, though (as the author points out, sometimes hilariously) less honest ones have only too easily fooled themselves and their colleagues. But dishonesty is not the main reason, or even an important reason, why Darwinism remains pervasive. It persists because it is useful to biologists, and because it gives every indication of being right. However, Michael Pitman has quite properly seized on the weaknesses that have always flawed the foundations of Darwinism, and he does us a service by worrying them to shreds.

What are the weaknesses? This book is waiting to tell you and I shall not steal its thunder, but two stand out above all. The first is that evolution by natural selection is remarkably difficult to demonstrate by the kinds of experiments that scientists normally accept in evidence. The second is that, while micro-evolution (the genesis of varieties of plants or animals) by natural selection seems plausible enough, macro-evolution (the genesis of widely differing patterns of organisms) by an extension and elaboration of the same process is much more difficult to swallow. Accepting the implications of macro-evolution by natural selection is, in the author's view, unwarranted on the evidence. Could the human eye have been evolved by the trial-and-error of natural selection? And the human brain?

Could the complex biochemistry of even the simplest bacterium have arisen by a series of chances? To believe that they could, says Michael Pitman, is just as much a matter of faith as belief in Creationism or any other less scientifically pretentious alternative to natural selection.

Is Darwinism, then, a scientific hypothesis or a religious experience? I have not found all the arguments of this book convincing, but I cannot disagree with its major proposition that, in measure of credulity, there is currently little to choose between Darwinist and Creationist. Michael Pitman may convince you (and if he fails, it will not be for want of trying) that not all Creationists are mindlessly conforming to outworn myths, and that there is more in the Creationist argument than scientists are generally willing to concede. He may even convince you, as he has convinced me, that some fundamental truths about evolution have so far eluded us all, and that uncritical acceptance of Darwinism may be counterproductive as well as expedient. Far from ignoring or ridiculing the groundswell of opposition to Darwinism that is growing, for example, in the United States, we should welcome it as an opportunity to re-examine our sacred cow more closely. Somewhere between us may lie the grain of truth that we would all be better for knowing.

Bernard Stonehouse
Cambridge, 1984

1 Where are we from?

Man is unique in his ability to seek his own origins — the only being known in the world to have the necessary level of consciousness. The question 'Where are we from?' has been asked through the ages and given many answers, intuitive or mystical, simple or ingenious. Today in the scientific age we still ask, bringing a wealth of new information and insights to bear on the problem. The question is a simple one but its implications are profound, for man's image of himself is enshrined within the answer. Did we evolve by chance, as part of the general process of evolution that many believe has created the universe? Or are we special, not only in consciousness and animal physique, but in the way of all life — set deliberately apart on a different level of creation from the rest of the inanimate universe?

This book is an attempt to marshal some of the facts that bear on the question of origins; it does not set out to answer the question, but the reader who stays with me to the end should find himself well equipped to provide his own answer. Throughout the book many different lines of evidence are invoked and many philosophies touched on. The name 'Darwin' appears frequently, though this is by no means a manual of Darwinism. Creationist views are given a fair hearing. Is this 'anti-scientific'? Science is a method, powerful but limited in scope to what its instruments can perceive. It cannot properly address certain questions — for example, those which concern unique historical or prehistorical events. Yet we may want answers. This is not 'anti-scientific'.

Furthermore, a rationalist regards a body as a machine with functional parts like pumps, biochemical engines, waste-discharge systems etc. It is not considered 'anti-scientific' to regard a machine like a television or a refrigerator as a creation, so why a body? Indeed, the creationist viewpoint is entirely mechanistic except with respect to the origin of the machine. For clear contrast with the Darwinian argument, I play devil's advocate for the modern creationist synthesis. Religious views are taken into account, though my own religion (if any) will not be apparent. My objective is

13

simply to disentangle the dozens of threads of argument that interweave around the questions of man's origins, and provide a new starting point from which they can be followed — separately and jointly — in a more orderly way.

One of the earliest recorded views of man's origins was due to Anaximander, a Greek philosopher of the sixth century BC. His theory of evolution taught that the earth was condensed from water to mud, from which arose plants and animals. Humans later arose in the sea, metamorphosing from creatures with fishy skins to walk on dry land. But the modern approach to man's history may be said to have started with Buffon (1707–88), a member of the French Academy, who was probably the first person to study nature in order to create a general theory of evolution. Buffon recognized that geological strata represented stages in history; in this he slightly anticipated the uniformitarian approach to geology that later came to be developed by, and linked with the name of, James Hutton (1726–97).

The principle of uniformitarianism is contained in the maxim 'the present is key to the past': the geological processes that have formed the earth's surface and its underlying strata are no more nor less than those we see operating about us today. This view, reasonable though it must seem in modern times, was controversial to the point of blasphemy when Hutton advanced it. Uniformitarianism, which invoked the slight but inexorable forces of rain, frost, wind and running water to explain strata, opposed the currently held view that layers of rock represented successive catastrophes — violent episodes of earth history of which the biblical flood was a prime example. For Hutton the forces of nature ruled; for catastrophists it was an intervening God, and Holy Writ supported the catastrophists. Modern geologists draw cheerfully on both views without the complications of religious explanations. Catastrophes did and do occur, interspersed locally among the uniformitarian processes that are mainly responsible for the layered structure of the earth's crust. It is becoming increasingly apparent that large-scale upsets have from time to time wrought havoc with the earth's geological form and with the plants and animals that inhabit the earth; catastrophes may well have influenced the course of world history far more than was generally recognized by the uniformitarians whose zeal reformed geological thinking in the eighteenth and nineteenth centuries.

Erasmus Darwin (1731–1802), a contemporary of Buffon and Hutton and remarkable grandfather of Charles Darwin (1809–82), was among the first in England to set out the principles which underlie modern theories of evolution. His ideas (vigorously expressed in his two books *Zoonomia* and *The Temple of Nature*) were intuitive rather than scientifically based. Their main contribution was to keep alive the spirit of controversy which faltered, as always, under the weight of orthodox religion. Had not Bishop Ussher, through genealogies leading back to Adam and Eve, calculated that creation began on the night preceding Sunday, 23 October, 4004 BC?

Charles Darwin, who never knew his grandfather let alone Bishop Ussher, drew on other sources that arose more directly from science. Most celebrated of all who have contributed to the question of man's origin, he did so at first reluctantly. There is little mention of man in his most famous book, *The Origin of Species*, which first appeared in 1859. Although he realized that man's physique resembles that of apes, he also realized the controversy such an observation would cause. His work was much influenced by Hutton's uniformitarianism, as recently popularized (in the 1830s) by his friend, the prominent geologist Sir Charles Lyell (1797–1875). In 1863 Lyell published his own major work *The Antiquity of Man*. Darwin followed in 1871 with *The Descent of Man* and was happy to confess that many of his ideas 'came from Sir Charles Lyell's brain'.

Thomas Henry Huxley (1825–95), zoologist, popular writer and lecturer, took up the controversy on Darwin's behalf, precipitating in his anti-clerical zeal a further controversy – between religion and science – that sadly obscured the real issues and still bedevils discussion of man's origins today. Like Darwin, Huxley was convinced by the fossil record and the long time that would undoubtedly have been needed to establish it, and had no inhibitions about disposing of Ussher. His quarrel was with an authoritarian church rather than with God. An agnostic (he coined the word), believing that the existence of God and the spiritual world could not be proved, he nevertheless recognized that a First Cause could not be ignored and defined God as 'a being absolutely infinite, that is, a constant substance with infinite attributes'.

Huxley's controversy with churchmen is epitomized by his famous exchange with Samuel Wilberforce, Bishop of Oxford, at the Oxford meeting of the British Association in 1860. The Bishop, a charismatic figure known popularly as 'Soapy Sam' who was much in demand as a speaker on contemporary matters, was spokesman for those who found the idea of animal origins for man wholly repugnant. He announced in the course of the debate that, whatever certain people believed, he for one would not look upon the monkeys in the Zoological Gardens as connected with his ancestors, and inquired of Huxley whether it was on his grandfather's or grandmother's side that he was related to the apes. Accounts of Huxley's retort vary, but it was to the effect that he would rather be descended from an ape than from a bishop. Such pleasantries hardened attitudes, provoking a less than rational response to Darwin's idea of human evolution from the religious majority of thinkers throughout the western world.

Huxley took it upon himself to proclaim the theory to the public through a series of articles and lectures. From 1881–85 he served as president of the Royal Society. His championship was so successful that his grandson, Sir Julian Huxley, could aver that Darwin had removed the whole idea of God as the Creator of organisms from the sphere of rational discussion.

Darwin's View

What was Darwin's contribution to the question of man's origins? What did he say that was different, and why did the views of this mild gentleman invoke such hostility? He did not, as we have seen, originate the concept of general evolution or of human evolution in particular. His two-fold contribution was firstly to marshal evidence that species – once believed immutable – change with time, and to identify a mechanism called natural selection. His impact arose from the thoroughness of his work, the power of his writing, and its timeliness; Darwin was one of those men who, to an age grown ripe and fit for it, project an idea with irresistible power, simplicity and magnetism. He gave evolution grandeur. The closing lines of *The Origin of Species*, stressing both the horror and the majesty of his concept of natural selection, are a good example:

Thus from the war of nature, from famine and death, the most exalted of which we are capable of conceiving, namely, the production of the higher animals, directly follows. There is a grandeur in this view of life, with its several powers, having been originally breathed by the Creator into a few forms or into one; and that whilst this planet has gone cycling according to the fixed law of gravity, from so simple a beginning endless forms most beautiful and wonderful have been, and are being, evolved.

That his writings include matters now judged wrong or open to question does not detract from their value. Darwin lacked information on many topics that we now regard as crucial to an understanding of evolution. He knew little of cell structure, physiology, genetics, biochemistry or molecular biology, and his knowledge of the fossil record was fragmentary compared with our much wider experience and understanding of it today.

What did he say? Essentially, he said that, within every species, many more individuals are produced than can possibly live. Limitations of food supply, predation, disease and other natural controls intervene to reduce the surplus, and there is thus a struggle for existence between members of the species. In this struggle the weaker members die and the fittest survive. 'Fitness' is made up of the small variations or departures from average that all individuals show; particular combinations of variations give some individuals advantage over others, and these are the individuals that will survive to form the next generation.

In this way natural selection (which Darwin compared with the process of artificial selection by plant and animal breeders) ensures that favoured variations will be well represented among the offspring of survivors, while unfavourable variations will remain at a low level, decrease or die out. Varieties, said Darwin, were incipient species; over the course of geological time new species would arise from old ones, often replacing ancestral forms by the process of natural selection.

Darwin's concept was, incidentally, an attack on biological orthodoxy as represented by the creationist Carolus Linnaeus (1707–78). Linnaeus, a Swede, was a taxonomist or classifier of plants and animals, whose system of naming and listing living organisms was (and still is) in widespread use. Its author had no intention that his system should be interpreted as providing evidence for evolution between species or, in a grand overall scheme, as a tree linking all life to common ancestors. Wrongly, in the view of modern creationists and evolutionists alike, he regarded species as fixed entities with identifiable, immutable characters.

For Darwin the species was not fixed. Although it could be listed and classified, it was an arbitrary and subjective entity based on a comparison of bodily features (the phenotype). He catalogued evidence, covering fields as wide-ranging as palaeontology, animal behaviour, pollination and pigeon-fancying, to support the idea that, by a continuous gradation of form, one species could evolve by natural selection into another. Darwin's title for his book is curiously misleading. What is a species if not quite a species? Because his definition was arbitrary and blurred, he actually sidestepped his title issue. If the title of the work was, in this respect, anomalous, so also was his use of the word 'Origin'. What he in fact meticulously and correctly documented in his work was what modern evolutionists call 'micro-evolution'. This occurs all the time. It is what creationists call 'variation' and is not support for evolution in the grand sense of the word. 'Macro-evolution' – the origin of major groups of organisms from common ancestors – remains a problem, as we shall see in later chapters. Few modern biologists feel that macro-evolution is just micro-evolution writ large; different processes altogether may be involved, and here Darwin's strength fails.

A notable weakness in Darwin's personal store of knowledge concerned heredity: though Gregor Mendel's work on inheritance was published in 1866, Darwin (like most of his contemporaries) knew nothing of it and half a century would pass before Mendel's experimental results and interpretations became widely known. Darwin was well informed on the practices of animal and plant breeding, and knew that some characters breed true while others are eclipsed; he knew also of mutations and their value to breeders, though he doubted if 'sports' (as he called them) had much effect in natural-selection processes. But neither he nor others who entered the debate on evolution had a model of genes and chromosomes, of linkage and segregation, of dominant and recessive characters – of the actual mechanisms of inheritance – to draw upon in their discussions. Thus Darwin's theory of evolution by natural selection was weakest where it should have been strongest. It lacked a theory of inheritance – a near-fatal flaw that, by the end of the nineteenth century, had almost caused the flame of Darwinian evolution to putter out.

What about the origin of life itself? It used to be thought that life could be generated spontaneously. It was a 'fact' that insects rose from mud and slime, and rotting meat bred maggots and flies. In 1668 Francisco Redi proved this was untrue but later, when scientists discovered bacteria and other micro-organisms, the problem cropped up again. In 1860 Louis Pasteur finally demonstrated that organisms are *not* spontaneously generated from a sterile environment.

It is clear, however, that where there is no spontaneous generation there may lurk a Creator. Darwin seems to have allowed for the special creation of at least a prototype cell: but then, again, in a letter written in 1871, he favours the view that life might have arisen from non-living matter in the past, without divine intervention: 'But if (oh! what a big if!) we could conceive in some warm little pond, with all sorts of ammonia and phosphoric salts, light, heat and electricity etc., that a protein compound was chemically formed ready to undergo still more complex changes.'[1]

To whatever extent they believe them true, Darwin's ideas have proved an immensely fertile source of inspiration and guidance for biologists in every field of research or inquiry.

The New Darwinism

Neo-Darwinian theory – Darwinism as biologists understand it today – can be summed up in the form of two propositions and a corollary:

(1)   There has been chemical evolution from non-life to life.

(2)   A common ancestry (phylogenetic relationship) is assumed to account for similarities of anatomy, physiology, biochemistry and behaviour that occur between different kinds of living organism.

It is therefore axiomatic that any one type of organism has been transformed from another kind, by either a gradual or a 'jerky' process involving random mutations acted upon by natural selection.

Though these statements form part of the background of every biologist – indeed of every biology student from junior grades upward – it is worth noting that neither of the propositions is an established fact, and the corollary, however axiomatic, has no strength of its own beyond the propositions on which it is based. This trio of ideas has, however, produced several decades of progressive biology. The concept of evolution has broadened far beyond biology; as Sir Julian Huxley (1887–1975), grandson of Darwin's 'bulldog' and an eminent biologist in his own right wrote in J.R. Newman's book *What is Science?*:

Evolution in the extended sense can be defined as a directional and essentially irreversible process occurring in time, which in its course gives rise to an increase of variety and an increasingly high level of organization in its products. Our present knowledge forces us to the view that the whole of reality is evolution – a single process of self-transformation.[2]

However, while most biologists remain evolutionists, many have reservations about Darwinism and neo-Darwinism, and their grounds for doubting the efficacy of natural selection are far from negligible. For P.-P. Grassé, renowned French zoologist, who is a former President of the Academie des Sciences and editor of the thirty-five-volume *Traité de Zoologie* (1948–72), the only scientific approach to evolution is through palaeontology, the study of fossils. He rejects Darwinian theories of the mechanism of evolution as inadequate: 'Directed by all-powerful selection, chance becomes a sort of providence which, under the cover of atheism, is not named but secretly worshipped. . . .'[3] Another critic, Professor Niles Eldredge of the American Natural History Museum, is more laconic but no less damning. In the *Guardian* of 21 November 1978 he wrote that the smooth transition from one form of life to another, implied in Darwin's theory, is not borne out by the facts. Lack of 'missing links' or transitional forms between groups of living creatures (for example, apes and man) has for long been attributed to an imperfect fossil record, but the past decade has brought to light strata representing all divisions of the last 500 million years, and no transitional forms have been contained in them. 'If it is not the fossil record which is incomplete,' concludes Professor Eldredge, 'then it must be the theory!'

To complete this preliminary examination of Darwinian and neo-Darwinian evolution, let us examine more closely what the word 'evolution' means and how it is used in a biological sense. Let us then see to what extent it has been, and can be, examined scientifically, for there are those who may be prepared to accept evolution — like God — as an article of faith, while others require the proof that its status in science demands.

What is Evolution?

The word 'evolution' means a process of development or unfolding of potential. We can speak of the evolution of a star, implying its genesis, development and destruction — the story of a spectacular bonfire far away across the universe. We speak also of the geological evolution of rocks and rock formations — the forces that form rocks and tear them apart, regenerate them in new forms and break them down again in a seemingly endless cycle. In chemistry we refer to the evolution of a gas. In common speech we use it of technological, social, political or other plans that are developed and put into practice; we may even speak of the spiritual evolution of a person — the process by which he strives to free his soul from its fetters of mind and body, turning consciously from evil and growing toward what is good. In biology evolution expresses the development, over a long time-span, of complex organisms (including man) from simpler ones.

However, 'evolution' in biology also expresses the phenomenon by which, in a given environment, organisms that are well adapted to the conditions of that environment develop from forebears that are less well adapted, and much of the confusion in evolutionary thinking arises from the use of one word in these two distinct senses. The first process — the development of complex organisms from simpler ones — we have already called macro-evolution. The second, its smaller scale counterpart, is micro-evolution. As we have seen, Darwinian theory of natural selection applies clearly to the latter, though not everyone is satisfied that the case even for micro-evolution by natural selection has been established beyond doubt. More doubtful still is application of Darwinian theory to macro-evolution; there is simply no direct evidence, from palaeontology, biochemistry, embryology or elsewhere, that fish have been transformed into birds, bacteria into jellyfish or reptiles into whales. There is, however, much circumstantial evidence of common ancestry, and on that the evolutionist's faith in macro-evolution is maintained, despite the criticisms of contemporary creationists to whom macro-evolution is 'the transformist illusion'.

Is Evolution Scientifically Valid?

Though Darwin's theory of evolution by natural selection has long been the concern of scientists in many disciplines, there are those who claim that it is not a scientifically based theory. Creationists join with some evolutionists in doubting its scientific validity; Sir Karl Popper, the distinguished philosopher of science, has dubbed it unscientific in his book on scientific methodology, *The Unended Quest*. He calls it a 'metaphysical research programme' because it does not meet his most searching criterion: a theory is scientific only if it can in principle be falsified by experiment, and is capable of refutation.

Popper's test owes nothing to the subject of the theory under review; it is concerned only with testability. This is best seen from examples. The 'law' of biogenesis, which maintains that life can be produced only by existing living forms and does not arise spontaneously from living matter, is theoretically capable of falsification, because a single substantiated example of life arising from non-life would suffice to destroy it. The theory can therefore claim to be scientific. A claim that Neanderthal man suffered from arthritis can be tested scientifically, because arthritis leaves marks on the skeleton which are subject to scrutiny. A claim that he suffered from epilepsy would not be susceptible to proof or disproof from skeletal remains. It could not therefore be judged scientific, at least until further evidence was presented, even though epilepsy is as much a clinical condition as arthritis. Astrology does not pass Popper's test, for the statements of soothsayers, though based on a systematic approach, are all too often untestable and irrefutable. Is Darwinian theory testable to a degree that Popper would approve?

Much of it is not, as Popper himself has been at pains to point out. The assertions by Darwinists, for example, that peculiar features of an organism are 'adaptive' (i.e., have been acquired by natural selection to promote survival) are seldom directly testable; even less subject to test is the key proposition in macro-evolution that, because organisms can be classified according to their resemblances and differences, the groups so identified must be related to each other by common ancestry. These are important issues among Darwinists; if their theory of natural selection cannot meet the generally accepted standards for scientific theories, it has no intrinsic advantages over the theories of the creationists – which nobody claims are scientific.

Popper has to some degree relaxed his position in a way that favours Darwinism, admitting (in a letter to the *New Scientist* dated 20 August 1980) that 'the description of unique events can very often be tested by deriving from them testable predictions or retrodictions'. It could, for example, be argued that the Darwinian model would be falsified if fossils of advanced animals were discovered lower in the earth's strata than those of their assumed ancestors; only a single human finger-bone discovered in authentic Devonian strata would topple a huge edifice of contemporary science, and set the whole world thinking along new lines. To this extent the evolutionary model itself may be judged scientifically valid, even though the Darwinian explanation of how it came about remains, for the time being, unproven.

Textbooks often mention evolution and Darwinism in the same breath. Recent doubts about the efficacy of Darwinian methods of evolution (such as natural selection) have led some biologists to uncouple alleged causes from the phenomenon itself. Even if Darwin was wrong, they argue, the phenomenon of evolution has occurred. However, if we have evolution without being able to properly explain its mechanisms, we are back to where we were in pre-Darwinian days. The idea is like a hollow shell, without substance.

An atheist believes that evolution is the result of chance. Theistic evolutionists believe God, having created the universe, let purposeless chance evolve life. A creationist, dismissing this hybrid view as absurd, contends that an intelligent creator creates complex machinery, such as a living body, deliberately.

In fact, by rigorous standards all three theories are metaphysical. This is because a theory of non-deliberate design (evolution) requires proof that no designer ever existed; a theory of deliberate design (creation) requires proof that a designer did exist. Theistic evolution, less logically, requires both proofs! But because the intelligence of a designer can be materially grasped neither in a Boeing 707 nor a bacterium, it is a matter of inference. Neither of the above proofs is scientifically possible because the field of science is limited to the material realm. And therefore each theory of origin is metaphysical.

Notes

1   Darwin, C., from a letter of 1871 in 'Some Unpublished Letters' ed. G. de Beer in 'Notes and Records of the Royal Society, London', 14.1 and quoted by Fox, S. W., *Evolution and the Origin of Life*, Dekker Inc., New York, 1977, p. 16 (footnote).
2   Huxley J., in *What is Science?* ed. Newman, J. R., Simon and Schuster, 1955, p. 272.
3   Grassé, P. P., *The Evolution of Living Organisms*, Academic Press, 1977, p. 107.

2 Two Pillars of Faith

If the earliest evolutionist was Anaximander (p. 14), creationism has been in the books since there were any. Another Greek philosopher, Anaxagoras (5th century BC), believed a teleological principle which he called mind brought order and harmonic motion into original empty chaos. Two and a half thousand years later Albert Einstein (1879–1955) felt much the same, using words that all but the most hard-bitten scientist would respond to: 'The scientist's religious feeling takes the form of a rapturous amazement at the harmony of natural law, which reveals an intelligence of such superiority that, compared with it, all the systematic thinking and acting of human beings is an utterly insignificant reflection.'[1]

Einstein, like Anaxagoras, was taking a creationist view. Creationism, like Darwinism, has changed with the years, following men's thoughts and fashions of thinking as they seek more nearly perfect understanding of themselves and the universe in which they are born. In Darwin's time creationism was almost universally Biblical. In the Christian world God was the Creator and two slightly different stories of Creation were told in the book of Genesis. Today creationists may take a broader view. Only out-and-out Fundamentalists hold a literal belief in these versions of creation; others may hold different opinions, or no opinions at all, on the identity of the Creator, but strong views indeed in the reality of intelligence that underlies creation.

That Darwinism became the mid-nineteenth-century alternative to orthodox creationism – that it was seized on by radical philosophers and thinkers in many fields as avidly as it was rejected by the conservative – is hardly surprising. Creationism at the time reflected narrow respectability, irredeemably the province of Church, State and establishment thinking. Evolution by contrast marked a renaissance of free thinking, and, above all, of rationalism – an august method of thinking and argument handed down from the ancient Greeks and Romans. Eclipsed during the Dark Ages, rationalism flourished once more with the dawn of science in the eighteenth century. In Darwin's day it was again in eclipse, at least among respectable Britons, and badly needing a surge of energy to brighten its flame.

Fundamental to rationalism is the philosophical materialism of Democritus, Epicurus, Lucretius and other radical thinkers of classical times; this same materialism appealed strongly to some scientists, social reformers and non-conformists of Victorian Britain. United in opposition to establishment self-satisfaction, rationalists drew together to form a powerful, progressive and, on the whole, peaceable protest movement, their protests directed against a rigid hierarchy of clerical conservatism dominating Church, State and education. Darwin was no rationalist, but his theory of evolution by natural selection was eagerly grabbed by rationalists and used, somewhat to his embarrassment, as a stick to beat all in authority who opposed science, social progress and freedom of thought.

The effects of Darwinism on one pillar of orthodoxy we have already seen (p. 15). Bishop Wilberforce spoke up for an almost unanimous clerical opposition. Though not all the clergy favoured his tactics in public debate, practically all were dismayed at finding God displaced from the centre of creation and chance − blind chance as it used to be called − occupying his throne. Perhaps more surprisingly, Darwin's mechanistic approach found an immediate response in Karl Marx (1818−83), the German-Jewish philosopher-in-exile, whose self-appointed task in London was to rewrite history in communist terms and prepare the world for revolution. Marx had already designed a political and economic system for an atheist world before Darwin's major work made its mark. He was not slow to recognize that Darwinism contained much that was pertinent to his own philosophy; '. . . this is the book', he wrote to his disciple Engels in 1866, 'which contains the basis in natural history for our view',[2] and he would gladly have dedicated his own major work, *Das Kapital*, to the author of *The Origin of Species* if Darwin had let him.

At Marx's funeral* Engels declaimed that, as Darwin had discovered the law of organic evolution in natural history, so Marx had discovered the law of evolution in human history.[3] With its denigration of non-material aspects of human life, and its mission to uproot tradition and destroy creationist concepts in men's minds, communism remains one of Darwin's strongest adherents; Marxists adhere to the 'word' with an old-fashioned faith that curiously matches the Fundamentalists' faith in Genesis. After 1949 when the communists took control of China, the first new text introduced to all schools was neither Marxist nor Leninist, but Darwinian.

---

* The only Englishman present at the graveside was Marx's young admirer, Ray Lankester FRS, who later obtained high office as Director of the Natural History Department of the British Museum (1898−1907), was knighted and is still remembered for his reclassification of the Talpidae (mole!) collection.

Mind and Matter

Said Bishop Berkeley, as he passed rationalist Thomas Hobbes in the street one day: 'No matter.'

Replied Hobbes: 'Never mind.'

This succinct exchange sums up an ancient philosophical divide. In physical investigation, matter predominates and life is lacking. In psychological investigation the subjective element (consciousness) predominates, and matter matters less. Biology, which investigates the properties of life-forms, falls uneasily between the two, and the frontier between creationist and Darwinian rationalist weaves close at hand. Both are happy to agree that the material universe is an automatic process, unless locally redirected by life into 'unnaturally' complicated systems such as hives, cities or spacecraft. Both also agree that man is a complex biological mechanism, powered by a combustion system that energizes two computers, each with a prodigious store for holding encoded information. The biological store is DNA, the micro-spiral contained in every cell that encodes genetic instructions (p. 53). The psychological store, which includes instinct and memory, is housed who-knows-where and encoded on who-knows-what.

Whereas a biological machine *might* be the inevitable product of chemicals and circumstance, to argue that it 'could have' evolved is not to say that it 'must have' or it 'did'. Creationist and rationalist agree that the world process and life-forms, with their DNA, brains or other parts, are material. But only the modern Hobbes claims that mind and memory are, being just properties of the brain, entirely material. For both him and the creationist information is carried as electrical codes in the brain; the artificial intelligence of a computer is a fair analogy. For the rationalist the analogy is complete. For the creationist it stops short; the brain interacts with a separate, immaterial (as far as present instruments can detect) and conscious entity, mind. The computer has no such mind, any more than a detailed scale-model of a man has. Even creator-computers which might, in the future, design their own progeny, will have no such mind. Yet these computers, made of matter, originated in mind. Before computers there was man, in the image of whose mind they were made. Is man, unlike computers, a product solely of matter? Or, like computers, a product of mind?

The judgement seat in the Court of Origins is narrow and uncomfortable. Try to be fair, and from one side you are accused of peddling poetic metaphors of biblical creation, and from the other you are knocked for scientific atheism. Nevertheless, there are good reasons why we should not take too seriously the rationalist's criticism of intelligent order in the cosmos.

First, look to your eyes. It remains a mystery how the light that enters them is translated, via electrochemical impulses of the nervous system, into the thing we call vision – the glorious mental experience of the waking state. This is a transition from matter to mind, and for the creationist matter is only the lesser half of the whole truth. That we happen to know more of matter than of mind is unimportant; it happens to be easier to study by scientific methodology. But prejudice over the nature of mind – whether, for example, awareness is wholly chemical or not – is important indeed. If creative intelligence *is* wholly material, are not soul or a Creator figments of erroneous calculation?

Given a hearing in the Court of Origins, the rationalist might argue himself out of mind in this way: 'It is an axiom of empirical science, self-evident when impartially considered, that any physical change must be causally determined by some antecedent physical change. The changes in the human brain, including those which are accompanied by consciousness, are essentially physical changes. All conscious processes must therefore be due to antecedent physical changes. There is no room for any such concept as mind or any such purely mental process as free choice or free will. Conscious processes must therefore either be introspective concomitants of the corresponding brain-processes or else they must themselves be generated by, and therefore really consist of, the accompanying brain processes. Thought is material.'

But could a set of physical changes, physically caused, possibly 'correspond' to such conscious experiences as 'seeing that an axiom is self-evident'? Could they set in train the conscious logical transition implied by the word 'therefore'? If the whole sequence of argument put forward by this particular rationalist were no more than inevitable spin-off from a chain of predetermined physical processes, it follows that he could not help saying what he did. His arguments, as reasoned arguments, would carry no weight. Why should we take the least notice of what he says?

Presumably Anaxagoras's 'teleological principle' and Einstein's 'intelligence of superiority' have more to recommend them than this exercise in rationalist self-defeat.

Design

'We do not believe in the theory of special creation because it is incredible.' In this way Sir Arthur Keith, a distinguished anatomist of the 1930s, echoed the rationalist feeling. But life itself is incredible, starting with every cell of every organ of every organism that Sir Arthur investigated.

'Every organism', wrote nineteenth-century German philosopher, Schoepenhauer, using words with which modern biologists will concur, 'is organic through and through in all its parts, and nowhere are these, not even in their smallest particles, mere aggregates of inorganic matter.' A cell may contain 100,000 million atoms, and they are atoms in specific order. It is the origin of such an incredible organization that interests the creationist.

It has been suggested that the chance emergence of man is like the probability of typing at random a library of a thousand books using the following procedure. Begin with a meaningful phrase. Retype it with a few mistakes and lengthen it by a few letters. Examine the result to see if the new phrase is meaningful. Repeat this process until the library is complete. There is incredibility enough to sink a ship on either side of this graphic argument.

At the height of his intellect man has discovered such principles as flight, heat insulation and the operation of lenses; all were widely used in nature long before he appeared (or was ever supposed to have appeared) on the scene. If we can refer to design in the works of man, can we not use the same term for nature? Why credit the plagiarist but scorn the original inventor?

But if nature's 'designs' evolved . . .? They are none the less designs; there is no evidence that, left to itself with whatever start it had over man, chance could evolve machines for work like men do, even the soft biological machinery of life. The creationist stops haggling over terms and looks for the designer. Through any but blinkered eyes the biological world shows clear signs of planning and order. It is not the order that constitutes a crystal, but a more complex order − the kind revealed in a developing seed or a growing embryo − the kind that, in any other context, we would unhesitatingly think of in terms of ingenuity and deliberate design. Darwin himself admitted design; thoroughly versed in Paley's creationist *Natural Theology* (1802), he considered design the most powerful of all arguments for the existence of reason in the universe.

The basis of biological design (again, creationist and evolutionist agree) is coding. The biological world is packed with intricatè, cooperative mechanisms that depend on encoded instructions for their development, functioning and mutual interaction. Were the codes designed, or did they evolve? Information theorists know that complex, meaningful codes do not occur spontaneously. Intelligent input is needed; spontaneity breeds randomness, and randomness destroys both order and meaning. Engineers too know that their designs are not the product of chance, but of that most powerful anti-randomizer − thought. The logic, says the creationist, runs strongly counter to the self-contradictory notion of chance-built design. No law of physics or chemistry has yielded a single principle of naturalistic innovation, information increase or functional integration. Left to themselves, things become more random and less tidy. The more complex the system, the more elaborate the design needed to keep randomness at bay.

Reason tells us to accept the straightforward argument and the simple interpretation. For creationists, teleology, the doctrine of design or pre-ordained purpose, has the merit of simplicity. They accept it with grace, and shamelessly flout the taboo that forbids biologists to acknowledge purpose and deliberate design in nature.

But is this scientific? Creation cannot be observed or investigated. Have creationists then rejected the methods of science? A few may reject science no less enthusiastically than some scientists reject creationism. Others argue that creation has two aspects. The acts of creation, like the acts of macro-evolution, are elusive; the products of creation – call them creatures or *creaturae* – are not. If you examine a machine (*creatura*), you may have difficulty in understanding how it was created. This will not stop you believing that it *was* made, and made to designs which, properly examined, may tell you much of the mind behind them. Does the machine work economically and efficiently? Is it readily repaired and reproduced? Does it do its job well? If so, there is evidence of good design and a praiseworthy designer. Why not apply the same reasoning to biological *creaturae*, says the creationist, and be prepared to accept that jellyfish, like jet engines, can be intelligently (and superbly) designed?

But, argued the philosopher Hume, the argument from design is illusory: should we exist if things were not as they are? Life has evolved because the conditions for it have happened to be present on earth. This criticism is unanswerable but empty; whether by accident or design things are as they are, and there are innumerable examples of design, all large as life, to balance against Hume's conjecture. Why should we take it seriously?

Design is a dead hand laid on research, claim the Darwinians; what is there to keep humans curious in a designed universe, where the answer to everything is 'God made it so'? Not at all, replies the creationist. Study the working of a Swiss watch, a spleen or a sympathetic nervous system – you'll be no less intrigued if you believe all three are products of purposeful design. Engineers study human artefacts with unending delight. Discussion or even acceptance of creationism need in no way reduce our capacity to enjoy biological research – once the mist of Darwinian myth has been dispersed. Indeed, if evolutionary theory were discarded today as inept, immoral or un-British, it would not change by one jot the mechanisms, processes and organisms that biologists found themselves studying tomorrow morning. Not every biologist thinks evolution all the time. Biochemists, for example, often find it useful to think 'as if' they were a cell's manufacturer, and ask themselves what programme they would develop to expedite a particular process. A trick of the trade? Yes, but good creationist thinking as well.

More about Coding

Several important biological discoveries since the Second World War have thrown light on the mechanisms of genetic coding. We know that nucleic acids DNA and RNA carry the genetic information, in the case of DNA through its celebrated, self-replicating double-helix structure. The genetic code has been satisfactorily cracked, and procedures have been developed for producing recombinant DNA (hereditary molecules) almost to order. Whatever made it, man now makes life − although he still does not understand even the general features of the origin of the genetic code that so precisely determines it.

Codes are a form of substitution, sometimes to hide things, sometimes to make them clearer or more convenient. The code of language uses combinations of sounds (or letters) to express ideas. Morse code substitutes dots and dashes for letters because they are easier to transmit by radio. Musicians follow a notational code − symbols on paper that keep them in order, allowing a hundred instrumentalists to play together in harmony and to repeat the performance, recognizably if not exactly, as often as they will. Ciphers are codes that obscure ideas so that only the initiated can understand them. Five letters in a particular sequence may instruct an operator − one who knows how to decipher them − to press the button that starts things moving. The same five letters in another order might tell him it's tea-time.

Genes too are encoded symbols, strung out in meaningful sequences along the nucleic-acid spirals; they encapsulate instructions for the chemical processes that keep life going. To the teleologist, who looks for purpose and believes in design, life is the manifestation of the encoded genetic instructions. As thought precedes action and plan precedes implementation, the coding held in genetic material precedes its phenotype or bodily expression. The genetic micro-computer has spatial economy and efficiency that awe the microchip engineer. As programmed punched strips guide the factory drill, so the encoded programme in the nucleus directly controls cellular metabolism. It is a tight code of control, tightly operated; there is little left to chance within a cell.

It is worth considering further the analogy of music, for this is something more familiar to the majority than genes and biochemical metabolism. Those who propose creation should reasonably be expected to say how they think it happened, and music provides a useful parallel. When a composer hums a new tune to himself, he converts an idea (a particular sequence of sound images) into energy (sound waves). If it sounds right, he can encode it in musical notation (the score). Now any other musician can decode the idea at will, and play the tune on a piano, trombone or violin. So where is the tune and what is its origin? There is one form of it encoded in notes on paper, but you won't learn much more of it by analysing the ink. There's another form of it coming from the piano, but you would be a fool to wreck the piano in search of it. You might hear the composer humming it, but don't shoot him and open him up; it is an idea you are seeking and, however good an idea, you will never find it in matter. Matter, as the new physics has it, is 'frozen' energy; Buddhists invert the image slightly and call it a vibratory illusion, but either way there is emphasis on the all-important alternatives to matter – energy and mind – that chiefly concern the creationist. The material states – solids, liquids and gases – are merely manifestations that we study because we can see, feel, smell, measure and quantify them. Being busy about matter is one thing; understanding what lies behind it could well be just as rewarding, though much harder to accomplish.

A prime tool of the new physics, with Einstein's special theory of relativity, is quantum theory; physicists use it for any problem concerning the microstructure of matter, for it describes the interactions of energy with matter. Particles confined within atoms (quanta) behave in ways that can be described by formulae, and essentially similar formulae are used to describe standing waves (the kind that occur when a guitar string is plucked). Quanta in fact are the link between matter and energy.

Understanding particles through a 'harmonic' principle (physicists call it an eigen-wave function) has led to important insights into atomic chemistry. The rhythms of the periodic table, which help us to classify elements and predict their chemical behaviour, are a function of the electronic structure of atoms and the vibrant energy of their quanta. The same energy can be linked through quantum theory with the generation of light and colour. Sound and light shows are an evocative use of vibrant energy – the energy of sound waves in air and electro-magnetic waves in space – designed and controlled to produce particular effects. Is the spectacle of life itself not pulsing entertainment on a grand scale, not of sound and light, but of chemical energies in harmonic oscillation?

It is interesting, at this point, to contrast a different way in which the classical Greeks also derived insight and inspiration from the same relationship of music with the inmost parts of mind and matter. Was it not the legendary Orpheus, son of Apollo and the Muse Calliope, who, with his incomparable gift for music, was able to strike a sympathetic chord of life not only in the living but the unliving rocks and streams and even the king of the dead! The principle of his harmony resounded at the heart of the Orphic religion, whose mysteries were probably practised at Eleusis near Delphi. Socrates, and maybe Plato and Aristotle, who had such influence over Islamic and European philosophy and science until the age of rationalism, were among the initiates. The doctrines of Pythagoras show close affinity to the Orphics and Pythagorean connections with the famous shrine at Delphi, sacred to Apollo, were also quite plain to antiquity. Apollo was the god of light, music and poetry, that is, a source of vibrant energy and song.

If we return to creation, one thing is certain; ideas need matrices or substrates to be expressed upon, and every musician needs an instrument – even if it is only his own voice. Man expresses his ideas primarily through his body and its many functions of voice, touch and expression. A creator uses the universe as raw material. Although its chemical 'notes' cannot be called music, any more than the noise of wind in trees or water through the rocks, matter is like an instrument or keyboard. Only a musician can exploit the potential of his instrument and create a work of art.

Similarly, an industrial technologist materializes his ideas using raw material extracted from the earth. Of course, his purpose-built machine obeys the laws of physics and chemistry but its grade of integration, both structural and functional, far surpasses that found in its raw materials. It is necessary to stress this point because here evolutionist and creationist differ most profoundly. Darwinians are determined that matter came first, and that mind has arisen from it through aeons of trial and error. The origin of nature was a big bang, not a Generator. For the creationist, mind came first and created matter for its instrument to play upon. There followed the music or, if you wish, biotechnology of life whose score was encapsulated in a chemical called DNA. In this way the work could be reproduced again and again without the composer's attendance.

Here, a common difficulty appears to block the path. If man made machines and a creator made man, who created the creator? It was Sir Isaac Newton who observed that a First Cause was not for science. Take salt. It is made of sodium and chloride ions. What are these made of? Protons, electrons and neutrons. And these? Any attempt to answer fundamental questions about nature leads to an infinite regress. Who can understand that? Why should the human mind resemble a creator's, any more than the matter we know resembles the 'structureless cosmic egg' in which a big-bang astronomer (like an Orphic cosmogonist) believes? Whether material or intelligent, can we demand a working model of ultimate reality? Or must we take nature as we find it, interpreting the facts as reasonably as possible?

At any rate, the idea of biological 'music' is dynamic. It implies that chosen 'dusts of the earth' were, like notes of music, orchestrated into forms of life; and that the composer exactly directed the chemical energy of these dusts in the form of a score, which today is called DNA. The British Museum in South Kensington was designed, by order of its first Director, Richard Owen, as a 'Cathedral of Natural History'. Perhaps, as a creationist, Owen would have approved more than his successors of the celebration implicit in a musical analogy of creation.

Notes

1   Einstein, A., *The World as I See It*, Flammarion, France, 1979, p. 21.
2   Meek, R. L., ed., *Marx and Engels on Malthus*, International Publishers Co. Inc., New York, 1954, p. 171.
3   Ruhle, O., *Karl Marx – His Life and Work*, The New Home Library, New York, 1929, p. 366.

3 Hierarchy

Naturalists who travelled with the early explorers to Africa, India and the South Seas brought back a problem. Their large collections of plants and animals, most if not all new to science, had to be named, catalogued and, if possible, slotted into some kind of quick reference system – a system of classification. Naturalists of a dozen nations were collecting: how could their collections be related to each other, avoiding multiple names in a dozen tongues?

About a thousand different kinds of plants and animals were identified in the time of Aristotle (384–322 BC). He classified the animals as those with red blood (vertebrates) and those without (invertebrates), and grouped the plants as herbs, shrubs or trees according to size and appearance. The basic system of names and classification that biologists use today, however, began with the work of Swedish naturalist Carolus Linnaeus (1707–78). He grouped animals and plants according to the arrangement to their parts (structure), and gave a distinctive double-name (bionomial) to each species or kind of organism. The first half of the name indicated relationship, like the surname of a family: the second half was peculiar to the plant or animal in question, like a forename, and both were based on Latin or Greek roots, making them international. For example, he called the house-cat *Felis domesticus*, the lion *Felis leo*. Common use of *Felis* showed that he recognized a strong affinity or family resemblance between the two: *domesticus* and *leo* distinguished one from the other in a language that any educated naturalist of the time would understand.

Today we still use the Linnaean system of naming, and contemporary classification is based on the principles that he established, though it begs a question of key importance. Just *how* are *Felis leo* and *Felis domesticus* supposed to be related to each other, and how is either related to *Felis lynx*, the bobcat, *Felis pardus*, the leopard, or any other *Felis* or cat-like creatures?

'How' in this sense is a small word with a profound meaning – indeed several meanings that tend to become confused as we advance with the classification system. We say that each of these animals is a distinct species (a concept we owe to John Ray, an English naturalist who died two years before Linnaeus was born), but that all three are similar enough to belong together in a higher group (a genus) called *Felis*. We go further and say that these cat-like animals are more distantly akin to dingos, dogs and wolves (three species of the genus *Canis*), and more distantly still to, say, stoats and weasels (*Mustelis*). To express these greater differences we place each of the three genera in a separate family – the Felidae, Canidae and Mustelidae. But because these creatures are all meat-eaters and have many features of anatomy and structure in common, we group all three families into a still higher category – the order Carnivora. So we build up hierarchies of animal and plant groupings that in some way express relationships, and at each level within this hierarchy we can ask the same question – how are these groups of organisms related?

Relationships in human affairs suggest common ancestors; sisters and brothers have parents in common, first cousins share grandparents, and more distant relations have great grand-parents or even more remote ancestors in common. We can show this kind of relationship in a family tree – a useful diagram from which the genealogist at a glance (and the lay-man with a little thought) can work out the exact relationship of any two individuals at any level. Can relationships between animals – cats, dogs and weasels, for example – be expressed in a similar way?

Linnaeus did not have such relationships in mind, because he believed species had been immutable since the day of Creation. Darwin, however, did. His contribution to biology was to point out, against the Linnaean schoolmen of his day, that a measure of variation does occur, that the species are not fixed. Modern creationists, in this sense, agree with him. For them Linnaeus is responsible for the confusion that now exists between 'species' and what they believe is the immutable unit of biological creation, the 'type'.

Species and Types

What is the difference between a species and a type, into whose origins we are enquiring?

A species is a recognizable kind of organism; one species is distinct from another and distinctness is an important aspect of the species concept. Human beings form a species; whether pink, brown or yellow, male or female, giant or dwarf, they are more like each other than any is like a horse or gorilla. Dogs form a species; dachshunds, collies and St Bernards, however different in detail, are clearly dogs, and in nine cases out of ten would not easily be taken for anything else.

But the tenth case is important. Wolves (to Linnaeus a separate species and still so by common consent) could easily be mistaken for dogs. Indeed, they sometimes 'mistake' each other and interbreed when the opportunity offers. Herring gulls and Lesser black-backed gulls, that in Britain form two clearly recognizable 'species', intergrade as we follow their stocks east or west around the ring of temperate latitudes in the northern hemisphere. Biologists call this a 'cline', but whatever it is called it throws doubt on the concept of distinction between species. Humans cannot interbreed with other animals but horses can breed with asses (giving mules) and lions with tigers (giving rise to tigons, if the father is a tiger, or ligers). Carrion crows breed with hooded crows forming interspecific hybrids that may or may not be fertile enough to breed in turn. And hybrids are even commoner in plants, though the same concept of species is applied in botany as zoology. Wheat can be crossed with rye and sugar cane with sorghum. Different species of grass may be cross-fertilized, and so may citrus fruit trees. A cross has been claimed between a radish and a cabbage, which both belong to the mustard family.

Some of these difficulties can be avoided if the definition of species is extended. Biologists who think in these terms require organisms to be recognizably similar in shape or form, and of very similar genotype, i.e., with almost identical genetic material in their chromosomes (a similarity that can be investigated by biochemical tests). They must also be reproductively isolated – prevented in some way from breeding with other closely related species. Reproductive isolation can be achieved by, for example, living on the other side of a mountain range from similar forms, by having different breeding seasons or stratagems for mating, or by developing anatomical or physiological peculiarities that make inter-breeding impossible or very unlikely.

The concept of species at first sight seems simple, but there are still difficulties awaiting those who try to define it more precisely. Individuals of the same species may be almost identical – like two ants or two blackbirds of a kind – or so grossly different they could be mistaken for different species. 'Morphs' or variant forms of the peppered moth can be almost jet black, almost white, or any of a range of shades between. The female marine worm *Boniella* may be three feet long; the male of the same species is a tiny creature, barely visible, that lives parasitically in her excretory system. The translucent jellyfish you avoid in summer waters starts life as a polyp – like a much simplified sea anemone – anchored to the sea-bed. Only with difficulty can polyps of a species be matched to their adult forms because they have so little in common. And what can we make of the species group of fruit-flies called *Drosophila willistoni*, a complex of over a dozen closely related forms in which occur 'sibling species' — forms that look exactly alike but are reproductively isolated (i.e., cannot breed together) and may be genetically more distant from each other than dissimilar non-sibling species that happily interbreed?

These difficulties do not invalidate the idea of species but they should make us aware that the simple approach is by no means foolproof, even if we consider only the organisms that we see around us today. Delving into the history of organisms brings another set of problems. The bones of sheep and pigs that we find in ancient middens are recognizable as species, but are subtly different from their modern equivalents: species can change with time, even a relatively short time. More problems crowd in when we try to think of much older individual fossils as belonging to species, for we know nothing of their breeding habits, how different they were from others of their group or for how long their stocks retained the characters by which we distinguish them as species. Nor, of course, can we ever examine or know their genetic constitution.

Many biologists have given up trying to define species in the classical terms given above — as distinctive groups of organisms of similar genotype that are reproductively isolated from each other; they seek instead some alternative concept for identifying and grouping organisms — an alternative with fewer inbuilt 'ifs' and 'buts' that more closely matches reality. One such concept is the 'type'. It is not a new idea, but new emphasis is being placed on its importance as a natural unit, especially by those who are looking beyond a neo-Darwinian interpretation of organisms and their origins.

Hybrids between horses and asses, or between dogs and wolves, are not unexpected. Hybrids between horses and dogs, or even between cats and dogs, would be as outlandish and unbelievable as centaurs, griffins or the mermaid of a sailor's fantasy. Cats and dogs, horses and dogs, fish and women, are different *types* of animals: dogs and wolves, horses and asses, are similar types. Types are set apart by their similarities of shape and structure: the differences are equally apparent at genetic or biochemical levels, and in behaviour as well. Each type is a cluster of forms or species (we can still call them species, without ascribing special — and questionable — properties of reproductive isolation to them). The clusters may change from time to time, but the types remain constant.

In 1957 Frank C. Marsh proposed a test to define the kinds of 'type'. He knew that in nature animal psychology, behaviour and coloration play a large part in determining whether copulation occurs, so he stated that artificial insemination would be required. It would, he conjectured, demonstrate relationships within a 'type' if the sperm entered the egg and the first division of the zygote was successfully completed. This purely chemical and physiological test would rule out instances when sperm entered the egg and instigated embryonic development, but the incompatible male chromosomes were later thrown out and further development ceased. Types cannot invade or alter other types. Interbreeding between species within a type may be possible; interbreeding between species of different types is unheard of.

A species can be described from a single individual. Indeed, it usually is, from a single specimen rather confusingly called the 'type' specimen (using the word in a completely different sense) that is especially labelled and handled with reverence in the vaults of a museum. 'Type specimens' are a constant source of trouble, for there is no guarantee that the first specimen of a species picked up and described is truly typical of the species: it may have been an aberrant individual, or one that in fact represents only a small, local group. 'Types', in the other sense of clusters or groups of species, cannot be described from any single animal or plant. They can only be described from the group of characters that the cluster has in common.

Curiously, the creationist concept of type defined here has parallels with an 'archetype' concept developed in the late eighteenth century by the German poet and philosopher, Goethe (1749–1832).[1] He proposed that all plants were derived from an archeplant (the *Urpflanze*), an idealized organism that contained the essence of all plants. Whether or not Goethe visualized the archeplant as a common ancestor, its significance rested in its possession of the characters from which all plants are derived. For a creationist the type, in this archetypal sense, is the supreme unit in nature, the product neither of taxonomy itself nor of evolutionary divergence. Within the sphere of the archetype he admits endless, irreversible variation due to sexual recombination (p. 61), mutations (p. 55) and environmental factors: outside it he admits no phylogenetic connection with other types.

Darwin too may be said to have thought in archetypal terms but his archetypes were of another sort. They were 'common ancestors', the trunk and main branches of the evolutionary family tree from whose loins sprang the species which, like twigs, we observe both in the fossil record and today. To the impartial observer, creationist and Darwinian archetypes are equally valid and equally elusive – both support useful theories but neither has ever been found.

The difference is important. Early hypothetical 'common ancestors' were species capable of unlimited divergence. They could generate both 'specialized' species which did not evolve further or became extinct; and other 'common ancestors'. For the evolutionist the basic unit of macro-evolution is the species. Micro-evolution of species grades over time intô macro-evolution. For the creationist such evolution is an illusion; he notes that genetics have demonstrated a limit to the amount of divergence that can occur to a kind of organism. For example, despite the artificially severe conditions geneticists have meted out to them, fruit-flies have remained as much fruit-flies as ever, and E-coli bacteria have remained E-coli bacteria.

For the creationist the basic unit of creation is the archetype: this is an idea first realized in one or a few similar forms. These will vary in space and time, but only within the unvarying theme of 'type'. Their family tree began and will finish within that 'type'. It is the type, not the species, which is fixed. Looked at in this way, to explain the origin of species is a trivial exercise which misses the main point. It is the origin of types (or archetypes) by macro-evolutionary relationship or by creation − both of which are unseen, hypothetical methods − which is the real, vexed question.

The creationist goes further. He views life in terms of archetypal themes which operate on four main levels, not just that of the *creatura* as a whole. These include the molecular, cellular and organic levels. Biology generally calls similarities at these levels 'homologies' (see below). Chapters 7 to 11 of this book are concerned, having identified archetypes at each of these levels, to incorporate their presence into a creationist view of nature. This can then profitably be compared with neo-Darwinian views. It involves using a computer anaology, viz. that discrete genetic programmes each incorporate a different permutation of archetypal subroutines. These per-mutations are capable of generating, without phylogenetic relationships, the diversity of biological forms we see today.

A problem remains to be solved. How can hierarchical relationships of types and species be represented? Not, say the creationists, by a family tree. To draw up a family tree we need to know precisely who the ancestors were. Neither modern species nor the fossil record tell us enough and it is unscientific (indeed, unsafe) to guess. Creationists prefer a 'Chinese box' arrangement of categories within categories (fig. 1): species cluster within types, and types that show similarities to each other are contained in larger boxes (or higher taxonomic categories). To the creationist the important point about this arrangement is that it carries no implication of common ancestry. It depends entirely on similarities that can be observed, measured and assessed, and on nothing else: it says nothing of unobserv-able, hypothetical ancestors that may or may not have existed long ago.

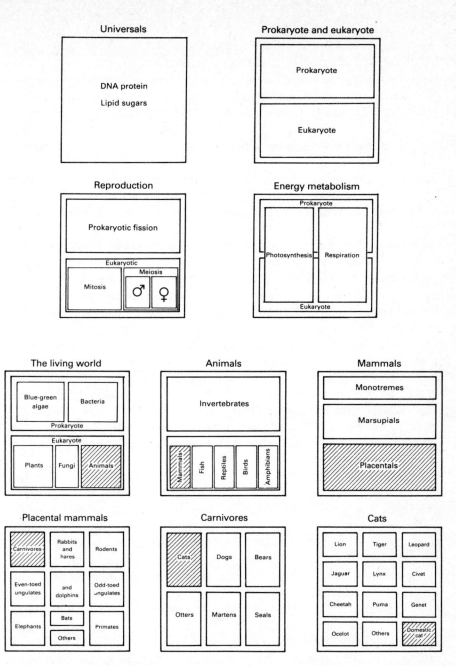

Figure 1. **'Chinese box' classification.** In this arrangement, relationships could be by descent or, like a range of motor vehicles, by design. Chinese box classification anticipates a new perspective in biology. The idea of 'complementary opposites', most sharply expressed in Taoist literature, is one developed in the later chapters of this book. This example pursues the classification of *Felis domesticus*, the family cat

Analogies, Homologies and Cladistics

Grouping plants and animals − whether into family trees or Chinese boxes − depends on an appreciation of similarities between them. The greater the number of similarities, the closer the relationship we believe to exist between the organisms possessing them. What are similarities and how do we assess them?

Birds, bats and bees show a striking similarity − they all fly. Are their wings similar? In function, yes, but not in structure, for birds and bats have bones in their 'front leg' wings, while bees have thin membranous wings held rigid by bonds of hardened protein. Bony wings, say biologists, are *analogous* to membranous wings: both work to press air downwards and hold their owners up, but there the similarity ends. Bat and bird wings, however, have a good deal in common. Not only are they bony; despite superficial differences of shape and covering, both are built on remarkably similar sets of bones that we can best compare by checking them off against the bones of our own arms − a single long bone (humerus) at the shoulder, twin long bones (radius and ulna) in the forearm and smaller bones that, with a little imagination, can be compared with wrist and fingers. This is much more significant for relationships, say biologists. Organs that are as much alike in structure as the bird and bat wing indicate a close relationship between birds and bats that is not shared by bees. However, a close structural similarity between bee and dragonfly wings suggests a close relationship between those two groups. Similarities of this kind are *homologies*.

Arthur Koestler described the presence of homologies as 'the preservation of certain, basic, archetypal designs through all changes'.[2] For example, the archetypal limb is supposed to have become adapted from fin to flipper, wing, arm or leg. Why, amid the flux of evolution, should an underlying pattern have remained so constant? What is the origin of such an archetype? Evolutionist and creationist agree on the significance of homologies but, as might be expected, they differ profoundly on the kind of relationship that homology implies.

To the evolutionist, homologous structures are clear evidence of common ancestry and a family tree of life. Bat wings, bird wings, flippers and human arms are similar because the ancestors common to birds, bats and humans had just such a structure − a forelimb built on the pattern that biologists identify as 'pentadactyl' or 'five-fingered'.

Creationists prefer to think of homologies as fixed patterns or discrete blocks, not unlike subroutines in a computer programme or pre-assembled units that can be plugged into a complex electronics circuit. They can be

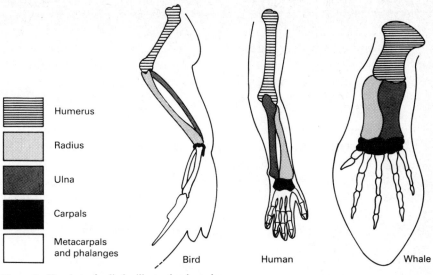

Figure 2. *Vertebrate forelimbs,* illustrating homology

varied according to an organism's need to perform particular functions in air, water or on land. Organisms are mosaics made up from such units at each biological level, and nothing of ancestry can be deduced from their possession. Grouping together animals with the pentadactyl limb, for example, tells us only that they happen to have that kind of limb in common – not that they inherited it from a common ancestor. Essentially, each archetypal part of every body is integrated into the service of four main interlocking themes of life – the generation of energy, reaction to stimuli, self-repair and reproduction.

Though unthinking Darwinians may accept homologies as firm evidence of evolution, thoughtful ones remain puzzled. Sir Alistair Hardy, former Professor of Zoology at Oxford University, wrote, 'The concept of homology is fundamental to what we are talking about when we speak of evolution, yet in truth we cannot explain it at all in terms of present-day biological theory.'[3]

An edifice of presumed relationships has been built up from the careful study of homologous structures – not least from the study of bones in the skulls of reptiles. However, does such study still support evolution? Recently a bone in the floor of the skull that lies between the eyes of amphisbaenian lizards has been causing trouble. In other lizards this bone, the orbitosphenoid, is formed in the normal way from a cartilaginous precursor. It was presumed that this also occurred in amphisbaenians and that the bone's unusual thickness was an adaptation to their burrowing habit.

Now it has been found that, in their case, it develops in the embryo in quite a different way – from soft tissue instead of cartilage. Because of this it fails the test for homology, although it completely mimics 'normal' orbitosphenoid bones in other lizards. Does it derive from a similar genotype? Could this kind of phenomenon have been more widespread than previously believed? An anatomist, Dr R. Presley of University College, Cardiff, has written: '. . . . this apparently obscure finding seems to me in the light of my present knowledge of the subject to have shaken the philosophical and logical framework of comparative biology to a very serious extent, and lots of people ought to be worried by it. I bet they aren't.'[4]

Similarly, the eyes of a mammal and a squid are superficially similar in appearance, remarkably similar in function and efficiency but built up in different ways from different elements. No Darwinian biologist would dream of 'homologizing' these two kinds of eyes, because they belong to quite different kinds of animals. He would just say that these eyes have 'converged' – become superficially similar, as fish and whales have 'converged' toward a similar streamlined shape that gives them equal mastery in water. Can homologous organs, then, be matched only in animals that we believe on other grounds to be closely related? If so, the Darwinist stands in danger of the noose of a circular argument; homologies cannot be based on relationships and, at the same time, be considered independent evidence for them.

Consider reptilian scales, bird feathers and fur. The evolutionist holds that feathers and fur have evolved, divergently, from scales. But can such different skin-coverings be called 'homologous'? For example, a feather and a scale develop from different layers of skin and follow different development paths; the feather's greater structural complexity must reflect a more complex genetic background. Yet the first known feather is entirely featherlike, not at all scalelike. The genes coding for each type of skin-covering must contain a sequence (subroutine) for keratin, because each is made primarily of a form of keratin. Yet this subroutine could well be integrated into quite a different overall set of genes. If so, how could we explain their origin in terms of simple inheritance from a common ancestor?

The pentadactyl limb, on the other hand, is more obviously a homologous pattern underlying the structure of wings in birds and bats, foreflippers and arms. It seems possible that broadly comparable groups of genes are involved in its production in different vertebrates. But does this necessarily imply a common ancestor? Aeroplanes, ships and motor cars all incorporate a 'homologous' steering system and, in most cases, a 'homologous' organ of locomotion. This organ, the piston engine, is adapted from a similar blueprint for use in air or sea or on land. In other words, the similarity of aeroplanes, ships and motor cars with respect to steering or motive power resides, basically, in a couple of archetypal ideas.

This is the way creationists approach the problem of homologies. For them the pentadactyl limb represents an encoded plan – an integral part of the overall encoded plan for vertebrate animals – that gives rise to basically similar structures in whatever animal the plan is realized. Alter the genetic coding in this direction, and up comes a seal's flipper; alter it in that direction and a bird's wing or human hand is formed. Who needs common ancestors? Similar subroutines could be inserted, amid quite different overall programmes, to make the similar eyes of an invertebrate octopus and vertebrate human.

An information engineer would recognize this process as one capable of being composed of elements he can understand. Computer programmes produce complex results from common elements and variables in much the same way. Understanding how it works, even if he could not devise such a programme himself, the programmer would not deem it, in principle, beyond the capability of an intelligence superior to his own – and he might laugh merrily if you suggested, in Darwinian style, that so complex a storage and retrieval system as the genetic code could evolve by a process of chance.

Derived from the Greek word for a branch, *cladistics* is a technique for assessing and tallying similarities or homologies. It does so without questioning what the homologies mean in terms of ancestry; it is entirely neutral, and a powerful tool in taxonomy. How does it work and what do we learn from it?

In classification organisms are collected into groups on the basis of shared features. Mammals, for instance, are all warm-blooded, hairy and, as regards their females, possess mammary glands. Some, like primates, have unusually large brains. Birds too are warm-blooded, but have feathers rather than hair, and relatively small brains. We can reflect these

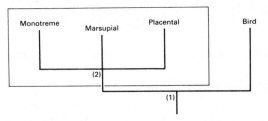

Figure 3. A cladogram, showing branched nodes (1) and (2). These nodes express *the proximity of relationships*. 'Mammals' are contained in a Chinese box

groupings of characters by a cladogram (branching diagram) whose nodes symbolize shared characteristics (fig. 3). Let node (1) represent warm-bloodedness and node (2) hair and mammary glands. Both mammals and birds share characteristic (1); only mammals share characteristic (2). The greater number of nodes two organisms share, the closer the link between them.

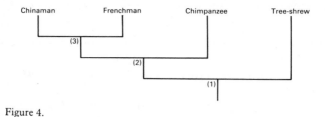

Figure 4.

Fig. 4 is a cladogram representing different kinds of primates, including man and the apes. The tree-shrew has a relatively large brain and eyes and is classed as a primate, though only a marginal one because it looks and acts like a long-nosed squirrel. It shares with other primates only the basic primate characters expressed by node (1). The chimpanzee shares homologies of 'advanced primates', such as binocular vision, with the Chinaman and Frenchman, but not with the tree-shrew. These characters are represented in node (2). The men share such homologies as bipedal gait and articulate speech (3) which are absent in the chimpanzee.

The more homologies each node is able to express, the more economically and accurately the cladogram expresses similarity; and the more closely linked are the organisms 'above'. Each node automatically includes all homologies expressed before it in the hierarchy, back to node (1) in the cladogram in question. Features shared by organisms can thus be arranged in a hierarchical pattern in nature. Node (1) represents the basic theme, whose variation is expressed in subsequent 'higher' nodes.

Data relevant for the nodes are furnished by anatomists, physiologists, molecular biologists, numerical taxonomists etc. The important thing to remember is that, although a cladogram is a kind of classification, its method is rigorously factual. Nodes represent the presence of characters: they do *not* represent common ancestors nor do branches symbolize phylogenetic trees.

The rigour of the technique forces the scientist to be objective and to sort out important from unimportant features of plants and animals.

Different cladograms can be drawn for the same organisms. For example, if in sorting through the primates we took into account the fact that both tree-shrews and Chinamen are found in China, ascribing this similarity to node (3) we would end up with fig. 5. But when other data were considered, we would find that homologies between Frenchman and Chinaman outnumber those between Chinaman and tree-shrew. The cladogram would be found wanting and replaced by one more economic-

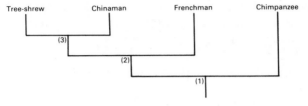

Figure 5.

ally expressing the sum of available data. Parsimony of expression is thus the main criterion of a good cladogram.

In this way information can be readily understood, criticized and tested. Cladistic relationships are discerned by clearly defined rules rather than whims or prejudice, and systematics is brought within the framework of scientific method; for these reasons, the technique now enjoys growing popularity among many of the world's taxonomists.

In no sense does cladistics deny evolution. It is a method for avoiding the circular argument wherein evolutionary hypotheses are used to erect a system of classification which is itself subsequently used as evidence that evolution has taken place. Evolutionists and creationists are served equally well by cladograms, though creationists, especially, find the method compatible with their ways of thinking. Advanced cladistics ('transformed cladists' are its exponents) are moving away from evolutionary theory, believing it prejudices the objective taxonomy they are trying to develop, and they are quite unconvinced that the neo-Darwinian mechanisms for evolution are adequate. As Colin Patterson has said:

As it turns out, all one can learn about the history of life is learned from systematics, from the groupings one finds in nature. The rest of it is story-telling of one sort or another. We have access to the tips of the tree; the tree itself is theory, and people who pretend to know about the tree and to describe what went on in it – how the branches came off and the twigs came off – are, I think, telling stories.[5]

Colin Patterson, a senior palaeontologist at the British Museum (Natural History), and those colleagues who share his views, have not necessarily disowned Darwin. But they have pointed out that, just as pre-Darwinian biology was carried out by creationists, so post-Darwinian biology is carried out by people whose faith is in the 'deity' of natural selection. They have set themselves to establish Darwin's theory upon unshakeable foundations by 'fleshing out', with trunk and branches from the existing twiggy evidence, the phylogenetic Tree of Life. This theoretical construct, however, has had little impact on progress in the work of biological research; and its assumptions remain unproven, even dubious. Systematics does not *need* an orthodox interpretation of the fossil record. Cladistics realizes that more than one church has been built on rock.

'If this is the result of cladistics,' writes Patterson, 'to give neo-Darwinism a good shake, then perhaps its critics are right to get excited.'[6]

Notes

1   Goethe, *The Metamorphosis of Plants*, 1790.
2   Koestler, A., *The Ghost in the Machine*, Picador, 1975, p. 142.
3   Hardy, Sir A., *The Living Stream*, Collins, 1965, p. 211.
4   Presley, R., private correspondence, April 1983.
5   Patterson, C., 'Are the Reports of Darwin's Death Exaggerated?', BBC Radio 4, 2 October 1981.
6   Patterson, C., 'Cladistics', *The Biologist*, vol. 27, no. 5, November 1980, p. 239.

4 Deep in the Womb of Nature

The womb of nature lies very close in all of us. There is no need to travel in space or time – only to follow James Watson, Francis Crick, Sydney Brenner and a galaxy of molecular biologists on the voyage of discovery that has carried them, over the last forty years, to the inner sanctum of the cell. Every cell is a time capsule. The seeds of the future are in it: the key to time past is there in a blob of cytoplasm far smaller than a pinhead. At the nuclear heart of each is coiled the DNA – the inspiration which engendered all that ever lived. You and I, each of us a trillion cells, are just transient waves – passing evidence of the power of DNA. To understand our origins, we must understand about DNA, the genes, the chromosomes that carry it and the cell that surrounds it and does its bidding.

Let us go back to 1860. Pasteur, a creationist, concluded experiments a year after the publication of *The Origin of Species*, showing that broth in sterile flasks did not spoil. Microbes from outside could not get in, and germs could not generate spontaneously inside the flasks although both broth and air inside were ready to support them. His work served to demolish the primitive notion of 'spontaneous generation', held at least since the time of the Greeks, and supported the law of biogenesis which holds that life comes only from living material of the same kind. This law throws no light on the origin of life-forms, although they must, like the universe itself, have had a beginning.

Not everyone was happy with Pasteur's results, or with their implications for life's origins. If spontaneous generation did not occur then life came to earth either from space (and how did it originate there?) or by an act of supernatural creation (excluded by definition from scientific investigation). Therefore, a materialistic explanation of origins had to be revived. However, applying artificial respiration to the dust of the earth puts a scientist in the awkward position of having to abandon a basic law – biogenesis – that has been universally validated and has no known exception. 'Only from a previous cell comes another cell': the principle of biogenesis states that a living thing can originate only from a parent or parents on the whole similar to itself. Huxley, Haeckel and Darwin found themselves in this fix in the mid-nineteenth century and solved it in various ways. Huxley and Haeckel seized at the myth of *Bathybius*, while Darwin dreamt of a prebiotic pond that doubled for the Fountain of Life.

Haeckel

Ernst Haeckel (1834–1919) was an avid, self-appointed spokesman for Darwinism in Germany, though his imaginative 'Darwinismus' expounded views far removed from the biological ideas or philosophical views of Darwin himself. Haeckel professed a mystical belief in the forces of nature and a literal transfer of the laws of biology to the social realm. The movement he founded in Germany was proto-Nazi in character; romantic Volkism and the Monist League (established 1906), along with evolution and science, laid the ideological foundations of National Socialism.

Of three main strands in his thought – German Romanticism, Materialism and Darwinism – only the latter concerns us. English Darwinism interlinked two main themes, natural selection and the struggle for existence. Social Darwinism is an attempt to explain human society in terms of evolution, but Haeckel's interpretation was quite different from that of capitalist Herbert Spencer or of communist Marx. For him a major component was the ethic of inherent struggle between higher and lower cultures, between races of men. There also existed a rejection of rationalism, and the embrace of *Blut und boden* (blood and body), man and nature, rootedness with earth. The form of social Darwinism in Germany became a pseudo-scientific 'religion' of nature-worship and nature-mysticism combined with notions of racism. This flared up into a full-bodied ideology with trends of imperialism, romanticism, anti-semitism and, as a racially united and powerful Germany, nationalism.

Emanating from the University of Jena, where Haeckel held a post from 1862–1909, his ideas held great sway. As a reputed scientist, he gave Volkism its respectable and appealing character. He studied radiolaria, sponges and medusae and wrote widely on zoological subjects. Having espoused the cause of Darwinism, he acted as the central European counterpart of Thomas Huxley. He spread the message widely. The three main planks of his argument, which he attempted to back up with evidence, were spontaneous generation, embryology and the wish to establish human ancestry with ape-men. At this point, only the first line of his argument concerns us.

Bathybius

In Darwin's time protoplasm was all the rage. Many a scientist was happy to agree that he could trace his ancestry back to a protoplasmic primordial globule of slime. The gap between living and non-living was one which the Darwinian evolutionist was obliged to fill. Haeckel proposed a hypothetical precursor to the amoeba, the moneran, 'an entirely homogeneous and structureless substance, a living particle of albumin, capable of nourishment and reproduction'.

The search was on. In 1868 Huxley, examining mud samples dredged off northwest Ireland some ten years earlier, identified a jelly in which were embedded tiny calcareous discs (coccoliths) which he incautiously linked with Haeckel's moneran. In honour of Haeckel he called it *Bathybius* ('life of the deep') *haeckelii* and, speaking before the Royal Geographical Society in 1870, maintained that it formed a living scum on the sea bed extending over thousands of square miles. There was great expectation in 1872 as HMS *Challenger* steamed out of port on an expedition to explore the world's oceans. However, no more *Bathybius* was found. Indeed, Mr Buchanan, the expedition's chemist, observed that he could produce the characters of the indescribable animal simply by adding strong alcohol (such as was employed to preserve biological specimens) to the mud. A specimen examined under a lens showed that calcium sulphate was precipitated in the form of a gelatinous ooze which clung around particles as though ingesting them, thus lending superficial protoplasmic appearance to the solution. Thomas Huxley's sample had been thus contaminated. Although it lingered in Haeckel's mind, for everyone else *Bathybius* died the death.

Modern Alchemy

In the 1920s a new wave of materialism swept Europe, and the origins of life came under scrutiny again. Not only was evolution crucial for the humanist point of view but so was abiogenesis − spontaneous generation, or the origin of organisms from non-living materials. After 1917 communism, an atheistic creed dedicated to the destruction of 'religious superstition', was applying itself to every aspect of life, including the origins, and the question of abiogenesis came under review. Could life have developed spontaneously without divine intervention?

Marxists Oparin (in Russia) and Haldane (in Britain) favoured the view that seemed most remote from religion and superstition; this suggested that simple organic compounds − hydrocarbons and even amino acids (the building blocks of proteins) – could arise spontaneously, and indeed had done so under the influence of such energy sources as lightning or radiation on the lifeless earth. Over long time-spans this material had aggregated into the much more complicated molecules – protein-like substances of high molecular weight, nucleic acids (DNA and RNA) and other compounds – characteristic of contemporary protoplasm. These large molecules of life are unstable in an oxidizing atmosphere, so their spontaneous formation required earth's primitive atmosphere to be without oxygen, as indeed independent evidence suggested it was. The Oparin-Haldane model assumed an oceanic 'pond' similar to that which Darwin had thought of in 1871. The complex molecules were supposed to have been acted upon, and changed within, this prebiotic medium, ultimately attaining the self-sustaining quality we call life. Abiogenesis was possible without special acts of creation. The earth amounted to a vast chemical factory in which life simply assembled itself.

Today billions of dollars are spent annually in attempts to demonstrate abiogenesis under laboratory conditions, by simulating conditions that are presumed to have existed on the primitive earth. Energizing 'soups' of simple organic materials causes them to combine into more complex molecules, but nothing approaching the highly complex molecules of living systems, which are each made of hundreds or thousands of correctly placed atoms, has yet appeared. Simple Victorian protoplasm has, under closer inspection, revealed a rich Elizabethan complexity which human ingenuity has so far scarcely begun to emulate.

The Cell

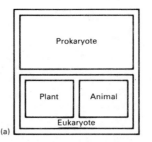

(a)

Figure 6. (a) Chinese box showing *cell types* (see fig. 1).
(b) A typical *prokaryotic cell*.
(c) *Eukaryotic animal and plant cells*, with organelles

Cells are the units of which the bodies of all organisms are made. The number of cells present ranges from one (in unicellular bacteria or protista) to fifty trillion or more (in you and me). On the basis of their detailed

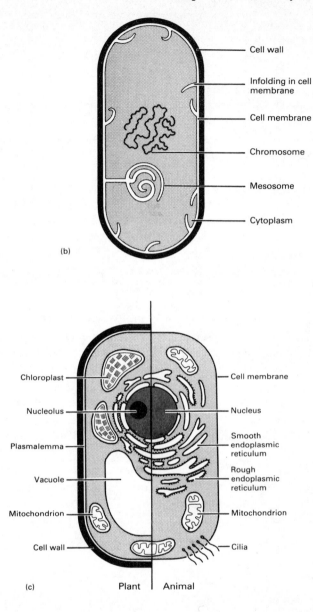

structure cells are classified as akaryotic (viruses), prokaryotic (bacteria and blue-green algae) and eukaryotic (all other types of cell). We will return to the status of the virus; for the purposes of evolution we are concerned only with prokaryotic and eukaryotic cells.

Prokaryotic cells (prokaryotes) are simpler in structure than eukaryotes:

they have no nucleus, mitochondria or any other of the internal structures (organelles) that characterize the cells of many-celled organisms. Their DNA is strung in a curious loop, not in strand-like chromosomes. They are a different kind of organism altogether − more different from the rest of living beings than, say, men are from trees.

Eukaryotes − common to men and to trees − fall into two main types. Those of plants usually have a cellulose cell wall around the cell and, when mature, a large fluid-filled vacuole inside; also chloroplasts, except in the strange case of a plant-animal protistan called *Euglena,* do not appear in animal cells.

The shape and detailed structure of plant and animal cells is very varied, and is correlated with the main function(s) that the cell performs. Such a correlation is found in any efficient, working machine: each part has a reason for its existence although, according to the evolutionist, parts of biological machinery were generated by chance. Prokaryotic cells, too, are complex though only a few millionths of a metre long. As well as a circular DNA chromosome they contain thousands of RNA and millions of fat, protein and polysaccharide (sugar) molecules, and their composition is dependent on the presence and correct position of these materials. In both prokaryotes and eukaryotes, the series of biochemical reactions which constitute metabolism must be carried out in the correct order: these reactions are facilitated by the presence of large, specific proteins called enzymes.

Normally the wind, sun and rain will quickly degrade large, complex molecules such as life requires. As water finds its own level, the natural tendency is towards chemical equilibrium; earth's tendency is not to produce proteins, DNA and other complex molecules, but to destroy them. Life is unstable, a disequilibrium which is maintained, as a juggler keeps juggling, by a continuous input of energy derived from sunlight (plants) or food (animals). In the sense that, for a time, it reverses the normal trend of things, life is decidedly unnatural − a joker that could never by chance have learnt the tricks of juggling. Even the bacterium is a much more competent juggler than chance alone could have provided.

Genes and Genesis

As we have seen (chapter 3), the basis of biology is coding. The code, written in DNA (deoxyribonucleic acid), is enshrined in every organism's genes and chromosomes − structures of which, you may remember, Darwin and the early Darwinians had no knowledge at all. In order to understand what is discussed in this and future chapters, it is necessary to get a clear idea of the structure and operation, as far as it is known, of DNA (fig. 7).

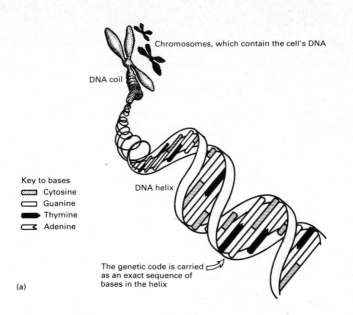

Chromosomes, which contain the cell's DNA

DNA coil

Key to bases
Cytosine
Guanine
Thymine
Adenine

DNA helix

The genetic code is carried
as an exact sequence of
bases in the helix

(a)

Figure 7. **Operation DNA – the inside information.** If we regard DNA, packed in chromosomes, as the book of life, then each time a cell divides to form a new cell the *whole* book is reprinted. In
protein manufacture *parts* of the book are translated into protein.

(a) *The structure of DNA*. In this diagram the structure is 'expanded' from a single chromosome in the
nucleus of a cell.

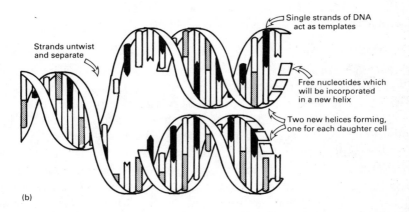

Single strands of DNA
act as templates

Strands untwist
and separate

Free nucleotides which
will be incorporated
in a new helix

Two new helices forming,
one for each daughter cell

(b)

(b) *The replication of DNA*. When a cell divides a duplicate of DNA is required. The two original
helical strands of DNA unwind; against each a new strand is built up, forming four strands in all. In
this way, each new double strand of DNA consists of half new and half old DNA.

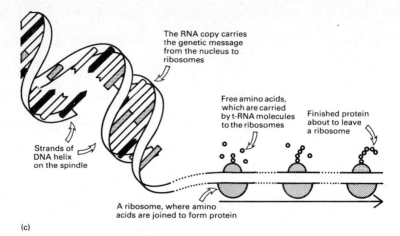

The RNA copy carries
the genetic message
from the nucleus to
ribosomes

Free amino acids,
which are carried
by t-RNA molecules
to the ribosomes

Finished protein
about to leave
a ribosome

Strands of
DNA helix
on the spindle

A ribosome, where amino
acids are joined to form protein

(c)

(c) *Protein Manufacture.* Here the process is slightly different. Limited unzipping exposes the
required sequence of bases along one DNA strand. These code for a specific sequence of amino acids.
Linked into a chain called a polypeptide, these amino acids constitute a specific protein with its
various important properties.

Specificity is the name of the game. Three DNA bases, a triplet called a codon, code for an amino
acid – for example, CGT for alanine. DNA makes RNA makes protein. First the DNA code is transcribed onto messenger RNA (m-RNA); m-RNA travels from the cell nucleus out to the surrounding
cytoplasm. Here there are transfer RNA (t-RNA) molecules whose job is each to collect and deliver a
single kind of amino acid to important little protein 'factories' called ribosomes. Each t-RNA
molecule has a specific 'anticodon' site which recognizes and binds to an m-RNA triplet corresponding to the amino acid it is carrying. Translation to protein occurs as m-RNA is fed, rather like tape
through a recording head, through the ribosome. T-RNA molecules on adjacent building sites on the
ribosomes unload their amino acids in order, according to the instructions carried on the m-RNA. The
product, as specific as any machine-tooled component, is a protein.

Modern genetics is the study of heredity and the unit of material
inheritance is the gene, often characterized as a short length of chromosome which codes for a particular polypeptide (protein) chain that
carries out a particular function or set of functions in the body. This is a
simple model of a very complex process. We know that single genes
sometimes affect a number of different characteristics (pleiotropy) – and
that several genes may interact to produce a single characteristic
(polygeny). Organisms are not patchworks with each patch controlled
by a single gene. They are integrated wholes, whose development and
maintenance is controlled by the entire set of genes (the genome) acting co-
operatively. Koestler summarizes: 'The gene complex and its internal
environment form a remarkably stable, closely knit, self-regulating
microhierarchy.'[1]

Chromosomes and the genes that make them up are composed of DNA.

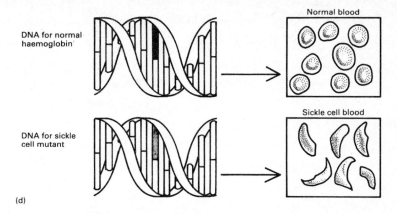

Normal blood

DNA for normal
haemoglobin

Sickle cell blood

DNA for sickle
cell mutant

(d)

(d) *Mutation*. A single error in the DNA base sequence for a protein can have major repercussions. In this case a single mutation has caused a severe illness − sickle cell anaemia. Distorted sickle cells get stuck in the tiny blood vessels of the body, so that their owner becomes starved of oxygen.

Dire consequences can result from other types of mutation. A prime characteristic of DNA is therefore its stability and fidelity as it copies and recopies the book of life. It encapsulates the drama of our human form. The zygote from which you grew weighs less than seven million millionths of a gram but contains all the necessary information for human structure. DNA is superpacked. The human population of the world is about four billions: DNA from the zygotes from which this multitude developed weighed less than 0.03 grams and could fit on the head of a pin!

A schematic diagram. For clarity, only one set of homologous chromosomes is shown.

This molecule was identified in the mid-1940s by Oswald Avery and co-workers as the chemical containing the genetic message in all animals, plants, bacteria and many viruses. DNA is in the form of a double helix, a duplex twisted like a two-stranded rope or length of electric flex. Each DNA strand is a polymer (chain) of nucleotides, and each nucleotide is made of three covalently linked parts: a sugar (deoxyribose), a phosphate group and a base. It is the sequence of bases along the strands that differentiates one DNA molecule from another, and provides the way in which it carries out its genetic functions.

There are four kinds of bases − guanine, cytosine, adenine and thymine. There are strict chemical and stereochemical (shapewise) rules by which bases may lie opposite to each other in the helix. Guanine on one strand pairs only with cytosine and adenine pairs only with thymine to produce the regular interlinked spirals of the DNA helix. These base-pairing rules are crucial for the biology of DNA replication which occurs in cell division and the production of gametes or sex cells (fig. 8).

When not involved in replication, some genes code for protein and RNA; some exercise control functions over the activity of other genes; and yet other, more repetitive sequences of DNA may turn out to be part of the

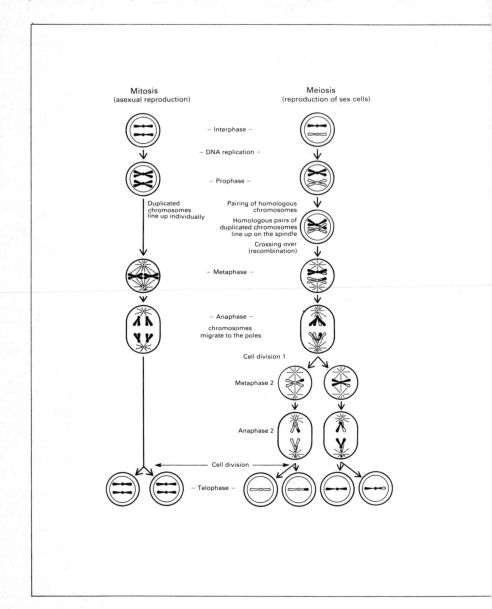

method by which the chains of bases are packaged to make up chromosomes. If we think of DNA as a reel of magnetic tape, genes correspond to permanent song-tracks recorded on that tape. The control genes, automatically and at exactly the right footage number, switch the right tracks on and off at the right times.

Figure 8. **Cell division – mitosis and meiosis.**

Mitosis. A cell (except a sex cell) reproduces by dividing into two daughter cells. Replication of DNA prior to cell division (fig. 7b) occurs invisibly in the *interphase* stage. Mitosis, the electro-mechanical dance in time and space by which each duplicate set of chromosomes is precisely despatched to its daughter nucleus, only occupies a small fraction of the cell's whole life-cycle. At *prophase* the chromosomes come into view as very long threads which shorten and thicken by coiling into a tight spiral. Having undergone replication at interphase, each chromosome appears as two 'sister' chromatids.

At *metaphase* the nuclear membrane has broken down. Each chromosomal pair of chromatids is attached to the equator of a spindle, which is itself slung between two points called centrioles at the poles of the cell. At *anaphase* the chromatids are pulled apart to opposite poles of the spindle. In this way accurate division of the genetic material is obtained. At *telophase* the chromosomes uncoil, elongate and disappear. So does the spindle, but a nuclear membrane is regenerated around each set of daughter chromosomes. The cell constricts across the middle until it completely separates the cytoplasm of the two daughter cells, each of which possesses exactly the same genetic constitution as the mother cell.

Meiosis. In meiosis the sex cells (gametes) are formed by double division. The main difference between meiosis and mitosis is that in the former homologous chromosomes associate and are then separated, giving the extra division. Ninety per cent of the cell's life-cycle is occupied with *interphase*. In *prophase* a homologous pair (marked white from the father and black from the mother), line up in what is called a bivalent. Each chromosome is seen to consist of a pair of 'sister' chromatids. Crossing over (fig. 9) occurs now. The nuclear membrane breaks down and, during *metaphase* and *anaphase*, each bivalent is split. The reassortment of either maternal or paternal chromosomes to each pole is random. In *metaphase 2* and *anaphase 2* the chromatids of each chromosome are split, as in mitosis. At *telophase 2* constriction occurs; the result of two cell divisions is that four sex cells, each with half the number of chromosomes present in the parent cell, have been formed.

Sex depends on this reduction. For example, each of your body's cells (except the sex cells) contains forty-six chromosomes. If in sexual reproduction both you and your partner contributed this 'diploid' number, the chromosome count would double at each generation (children ninety-two per cell, grandchildren 184 etc.). As it is, two meiotic divisions halve the count, generating 'haploid' sex cells with only twenty-three chromosomes. Therefore fertilization always regenerates the count of forty-six.

Meiosis, a triumph of biological engineering and the basis of sex, produces only eggs or sperm, and no other sort of cell

This is not all. For some extraordinary reason the genes of higher organisms turn out to be split into many pieces, quite unlike those of bacteria. The pieces are joined by stretches of genetic material that have nothing to do with the structure of protein for which the gene codes. The function of these stretches is unknown but may be regulatory. At any rate, stretches of m-RNA are snipped and spliced so that a relevant gene, which may be in several pieces, is reassembled to function as a whole.

There seems to be positive, purposive emphasis in nature upon preservation of the exact superspecificity of the DNA base-coding sequences. Every author knows that mistakes creep in as a manuscript is, in two 'generations', typed and then typeset. New bodies are generated from the sex cells: it may be argued that mistakes here are as inimical as mistakes in a text to its editor. Organisms have a battery of enzymes that detect and remove mistakes made in manufacturing their genetic library. For example, it seems that the enzyme DNA polymerase, which replicates DNA, may also be capable of proofreading and removing incorrect base-pairing. 'DNA polymerase has evidently evolved to a highly sophisticated functional state, almost as if it has a mind of its own'.[2] Of course it has not, any more than a computer has a mind of its own, but its quality for self-maintenance has a touch of magic that has impressed more than one sophisticated biochemist.

It is worth elaborating the analogy of DNA with a computer. This is done by Hoyle and Wickramasinghe who conclude:

'All this is strikingly similar to the situation in the living cell. For discs or tapes substitute DNA, for "words" substitute genes, and for "bits" (a bit is an electronic representation of "yes" or "no") substitute the bases adenine, thymine, guanine and cytosine.'[3]

The fundamental unit of data storage inside a digital computer, a bit, is either off (labelled 0) or on (labelled 1). A 'byte' is made up of eight bits and is capable of holding 2^8 (=256) different binary patterns – enough to represent such characters (letters of the alphabet, numbers, punctuation marks) as most users require. A byte can therefore stand for a character – just as in biology a 'byte' is a group of three nucleotides called a codon. This 'triplet' nature of the code, in which each codon stands for an amino acid, was first demonstrated by Crick and Brenner.

If we turn to the translation stage in protein manufacture (fig. 7), we can take the computer analogy further still. The code (DNA) is chemically dissimilar.from its product (protein made of amino acids). In living organisms there are twenty amino acids from which proteins are normally composed. For each of these there must exist a different intermediate (t-RNA) molecule to act as a specific 'link' between DNA and protein molecules, which have nothing else in common. The whole genetic business hinges around the presence and operation of t-RNA molecules.

As a computer computes in binary machine code, so the nucleus computes in DNA. But just as the body needs to interact with DNA, so the human operator needs to interact with the computer. He can do so because machine code is translated, by a compiler, into the 'alphanumerical' language he speaks and understands. The biological compiler, without which the body could not interact with DNA or vice versa, is t-RNA. How could DNA and body (or protein) language, together with their associated compiler, have arisen by accident, by mutation?

Like cards in a genetic pack, genes can be shuffled (fig. 9 overleaf). However, just as the most accurate electronic copier occasionally garbles a message, so a gene is occasionally garbled or a chromosome damaged. This is mutation (fig. 7d). As opposed to shuffling, it seems the only way that *new* cards can be added to the genetic pack.

Most mutations are changes of individual genes; that is, they involve loss, gain, substitution or rearrangement of bases in the DNA nucleotide sequence. Some, however, are gross structural alterations in chromosomes or changes in numbers of whole chromosomes per nucleus.

Whereas recombination and reassortment are built into the system, mutations are rare and discordant events. Clearly, only mutation in a sex cell has any value as far as evolutionary theory is concerned, because here it produces a source of variation that remains intact, as a heritable unit, from generation to generation. As well as through miscopying, mutations can occur as the result of ionizing radiation or treatment with certain chemical substances.

Because they seem to be the only way in which *novelty* can occur in a gene-pool, neo-Darwinian theory places great emphasis on random mutations. They constitute the raw material of evolution, the chance upon which the necessary force of natural selection acts. They are the rope, or the thread, upon which the theory of evolution hangs.

Now, armed with some contemporary knowledge, we can return to Darwin's predicament.

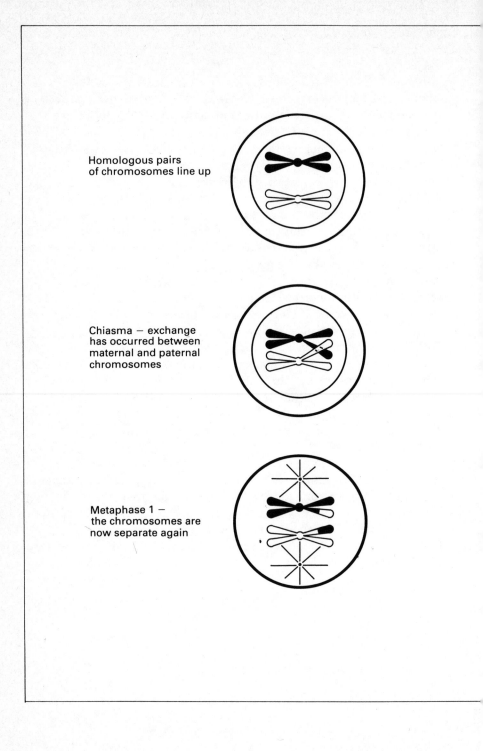

Homologous pairs
of chromosomes line up

Chiasma – exchange
has occurred between
maternal and paternal
chromosomes

Metaphase 1 –
the chromosomes are
now separate again

Figure 9. **Shuffling the genetic pack.** Laid into the meiotic programme for gamete production there exist two mechanisms, called *crossing over* and *reassortment*, which serve to shuffle the genetic pack.

Mother and father contribute an equal number of similar (homologous) chromosomes to a normal cell. In our case twenty-three maternal and twenty-three paternal make forty-six in all. During the early stage of meiosis, when these homologous chromosomes line up, the maternal and paternal portions exchange material. They do this in regions where chromatids, strands which result from chromosome replication and lie side by side until a later stage, appear to have broken and joined again, but to the 'wrong' partner. The process is called crossing-over.

Crossing over takes place during prophase of the first meiotic division. It occurs at the chiasma between chromatids of homologous chromosomes. Different genetic recombinations are formed where the chiasma (and therefore breakage) occurs between chromatids. The diagram is simplified and in practice several chiasmata may occur between different chromatids along the length of the bivalent.

How does crossing-over generate diversity? In fig. 7 you saw that a chromosome is composed of a double-stranded helix of DNA. Base-pairs of these strands make up the genetic code, operative sections of which are called genes. A gene is a short length of chromosome, influencing a particular set of characters in a particular way. Homologous chromosomes contain homologous genes, but it is believed that a minor difference in the DNA base sequence can cause these genes to express themselves differently, as alleles.

In simple terms, an allele is a variant form of a gene which governs a certain characteristic, for example, colour of eyes or hair. What crossing-over and, later in meiosis, reassortment does is to shuffle the alleles. In reassortment, homologous chromosomes are split up into four gametes whose content is neither exclusively maternal nor paternal but a random mix from each. This means alleles are further mixed. Crossing-over (recombination) and reassortment amount to powerful mechanisms which ensure that each gamete is the product of generous shuffling of parental genetic material. In this way offspring emerge as the expression of ever-fresh permutations of alleles. Each organism of each type is a unique expression of its own kind; more than this, such individuality guarantees that, in the long term, each kind of creature (or plant) will survive. Genetic shuffling acts as a buffer; it acts to ensure that in any population of organisms there will (almost) always be some members sufficiently adapted to withstand whatever changes occur in the environment. In this way crossing over and reassortment underwrite survival, as opposed to extinction. It is important, however, to notice that because they only shuffle the pack, they cannot be held to create any new material which might cause macro-evolution

Fleeming Jenkin

About half a century before Darwin the French evolutionist, Jean Lamarck (1744–1829), argued that new traits came from use and disuse of faculties, and from direct and indirect actions of the environment. For example, a trained boxer would produce sons (and daughters too?) genetically disposed towards large biceps. In the same way, lack of vegetation at ground-level might cause, by ancestral stretchings towards higher foliage, the gradual evolution of a long trunk in elephants, or neck in giraffes. Although this theory of 'evolution by acquired characters' is now discredited, Darwin (with some inconstancy) agreed with it: he could see no reasonable alternative.

It is clear, however, that the long neck of the giraffe could not have evolved without corresponding (and, in evolutionary terms, independent) changes in the vascular system: for example, the difference in blood pressure between the 'head up' and 'head down' positions could not be tolerated by the brain without an intricate system of valves which had to develop along with the neck. Moreover, the offspring of 'stretched' parents start off as disadvantageously small as all the others: we know of no mechanism by which a stretched neck can pass from one generation to the next.

Darwin shared with his contemporaries the idea that it could. They believed each organ secreted pangenes (or gemmules) which collected in the blood and flowed to the reproductive organs. A bigger neck would make more neck pangenes and so ensure the inheritance of its own acquired characteristics. But nobody could isolate a gemmule or demonstrate its existence under a microscope.

There existed the parallel notion, which Darwin in 1859 was much more inclined to believe, that *chance* variation might occur in genetic material, though no one was very clear as to what this might be. Variations that produced a modification, advantageous in a given environment, would be selected by nature. The mechanism by which parents passed on their stock of hereditary material was believed to be the 'blending of inheritance'. It was based on the natural enough assumption that the physical endowments of a new-born babe were a mixture (or blend) of characteristics of each parent, to which each parent contributed roughly half.

In 1867 Fleeming Jenkin, a Professor of Engineering at Edinburgh University, stepped into the picture. He published a review of *The Origin of Species* in which he demonstrated that any character, including a useful chance variation, when mated with another one, would, by blending, become diluted to 50 per cent of the first generation stock. It would be diminished to 25 per cent in the second generation and so on: thus the novelty would have disappeared long before natural selection could underline it. It was simple and it was obvious. No new species could arise from chance variations by the accepted 'law of ancestral inheritance by blending'.

Darwin, reportedly believing that this was the most valuable criticism anyone had made on his views, never saw how to get over the difficulty. There was an alternative theory of orthogenesis which said that evolution was pre-programmed and that each organism evolved along a certain line regardless of its environment or the reproductive consequences of each change. The notion of a master-plan, however, did not appeal to him. Reverting instead to an emphasis upon Lamarckian ideas, he inserted a new chapter into the sixth edition of *The Origin of Species* which resuscitated the idea of inheritance of acquired characteristics. Socially, this theory held: in some quarters it enjoyed a considerable vogue, still lingering until the 1960s in Soviet biology because such inheritance reinforced the lessons of the revolution and projected its children gloriously forward towards physical and mental utopia. Scientifically, it made no progress: scientific Darwinism in the late nineteenth century just chewed on its pipe. It awaited rescue, which ultimately arrived in the form of a neglected paper by an obscure and long-forgotten Augustinian monk.

Sorting it Out

Gregor Johann Mendel (1822–84) was a science and mathematics teacher at a high school in Brunn, Moravia. He was also a monk and, in the monastery garden, he carried out his famous experiments on the genetics of peas. He is called the 'Father of Modern Genetics' because, by keeping records of the results of crossing varieties of peas, he discovered the basis for the transmission of characteristics which holds true for all forms of life. In 1865 he presented his work before the Society for the Study of Natural Science in Brunn.

Mendel's Theory of Discrete Units of Heredity (later called by W. Johannssen 'genes') states that such units, as adamantine as the Newtonian corpuscle, combined into a great variety of mosaic patterns but did not relate or integrate in such a way as to lose their distinct, independent entity. Although they could be reshuffled, these 'atoms of heredity' were transmitted unchanged from generation to generation. Each simple factor determining a hereditary trait was contained in a Mendelian gene and each gene had its place, like different permutations of beads on a string, on the chromosomes of the cell nucleus.

Why was Mendel's paper, published in 1866, 'neglected' until 1900? R.A. Fisher, a founding father of modern biochemistry, wrote:

The journal in which it was published was not a very obscure one and seems to have been widely distributed. In London, according to Bateson, it was received by the Royal Society and the Linnaean Society. The paper itself is not obscure or difficult to understand; on the contrary, the new ideas are explained most simply and amply illustrated by the experimental results.[4]

In proposing a theory of inheritance through discrete, non-blending genetic particles, Mendel had unwittingly invalidated Jenkin's argument before it was even published. However, Mendelian theory raised a new spectre; if the particles were unchanging, from whence could information for new forms (and species) arise?

Remember what Mendel said. Inheritance depends on particles that are discrete; that is, they do not mix with one another. During cell division (mitosis) or in the formation of sex cells (meiosis) the 'atomistic' genes are shuffled; they undergo independent assortment but are not changed, either in kind or quantity. Mendel's theory allows any number of permutations to occur but the genetic information itself remains stable. Same pack of cards, no change. Mendel's work now seemed to rule out decisively the possibility of unlimited gradual change, such as Darwin's theory required. Nowhere on the voyage of the *Beagle* (after which he formulated his theory) did Darwin observe anything which forced him to the conclusion that change was unlimited. Had he known of Mendel's theory, it is likely he would never have published *The Origin of Species*.

In London, Thomas Huxley's 'X' Club,[5] an influential group of nine men who were notified of meetings by the delivery of an algebraic formula and who always dined immediately before meetings of the Royal Society, may have suppressed Mendel's work. The Club mustered a Secretary, Foreign Secretary, Treasurer and three successive Presidents of the Royal Society, six Presidents of the British Association and several officers of the Geological, Linnaean and Ethnological Societies.

It is certain that the 'gay and conspiratorial' 'X' Club, which was strongly evolutionist in character, not only influenced the appointments made for senior positions in the newly formed universities of the Victorian era but also, until its demise in the 1890s, practically controlled the business of the Royal Society. It never elected a tenth member: Busk, Frankland, Hirst, Hooker, Huxley, Lubbock, Herbert Spencer, Spottiswoode and Tyndall were its men. Is it possible they saw the implications of Mendel's paper and refused to confront them?

In 1901 the Dutch botanist, Hugo de Vries (1848–1935), concluded from experiments on evening primroses that a new species had arisen as a result of a sudden change in its germ plasm which he called a mutation.

In the course of developing his theory of mutations, he rediscovered Mendel's paper. Of course, his theory bypassed the difficulty which Mendel presented to evolutionists. In so doing, it has to this day rescued Darwinism. Although no satisfactory mechanism for evolution has been found, it is today generally accepted that an accumulation of successive, favourable mutations were acted upon by natural selection to 'create' all protista, plants and animals – most probably from a unique proto-cell.

To round off early theories of inheritance, we conclude with August Weissman who located 'germ plasm' in the chromosomes. He carried on experiments to see if characteristics acquired by the body (somatoplasm) of an individual could influence the germ plasm. For example, he cut off the tails of mice for twenty-two generations but found the twenty-third generation all had normal tails. His work caused the rejection of Lamarckian and pangenesis theories and helped establish the 'central dogma' of molecular biology, namely, that translation of the genetic code occurs only in the nucleic-acid-to-protein direction. In other words, germ cells are kept intact, unmixed with somatic cells. Messages go *out* from the DNA but none return. The DNA can, therefore, only be changed from within by mutation.

If critics inquire whether mutations are sufficient to generate new organs and organisms, they also query the sufficiency of the second main pillar of neo-Darwinism, the necessity which acts upon mutational chance – natural selection.

Notes

1   Koestler, A., *The Ghost in the Machine*, Picador, 1975, p. 134.
2   Holliday, R., 'Nature's Means for Preserving Genes', *New Scientist*, vol. 84, no. 1182, 22 November 1979, p. 599.
3   Hoyle, F., and Wickramasinghe, C., *Evolution from Space*, Dent, 1981, p. 106.
4   Fisher, R.A., 'Has Mendel's Work been Rediscovered?', *Annals of Science*, vol. 1, no. 2, 1936.
5   Bibby, C., on Thomas Huxley, *Scientist Extraordinary*, Pergamon, 1972, pp. 58, 135.

5 Sports, Survival and the Hone

Sports

Darwin ruled out 'sports' (now called mutations) as the source of novelty on which natural selection could work. After de Vries, though, neo-Darwinians resurrected mutations to play a key role, with natural selection, in the modern synthetic or neo-Darwinian theory of evolution (p. 18). Both concepts deserve close attention. Historically, Darwin's natural selection preceded its twentieth-century partner, mutations, but in biological practice it is mutations which must generate novelties before natural selection 'tests' their worth. So it is logical to take a look at mutations first. How probable is it that they measure up to the task eager evolutionists have wished on them?

Mutations can result from the alteration of just one of up to 50,000 atoms which are believed to constitute an average gene. They are naturally rare; one in ten thousand to one in a million may occur per gene in a single generation, though we can increase this rate by irradiation (with X-rays, gamma-rays, neutrons etc.) or by using chemicals like mustard gas or colchicine. The neo-Darwinian emphasizes that, from the standpoint of adaptation of the organism, mutations are random. There may be a tendency for certain physico-chemical bonds in the DNA to be broken slightly more easily than others, but mutations are random in the sense that we are unable to predict such an event or correlate its occurrence with the environment. Of course, X-radiation strikes, like lightning, at random. Mutation is *not*, for the neo-Darwinian, directed.

Most mutations are lethal or harmful; some appear neutral and a few (less than 0.1 per cent) can be interpreted as favourable. If the genetic blueprint for an organism is initially optional – like, say, the design for a new TV set – then mutations appear as damage incurred by wear and tear or misuse. Kicking a damaged TV set might improve its performance but the treatment is not generally recommended. In no way could random – or even well-directed – kicking have been responsible for the origin of the

TV in the first place. But the neo-Darwinian, who asserts that mutations are the raw material of evolution, and the only source of novelty for natural selection to work on, is both denying the existence of an optimal genetic blueprint (or archetype) for a life-form, and accepting 'kicking' as a rational means of improving it out of recognition.

Mutations are pathological aberrations: who would risk exposure to radioactivity to generate a superhuman child? Moreover, since polygeny and pleiotropy (p. 54) account for the closely knit functional integration of the genome, the probability of simultaneous good mutations in all the genes which control a given character must lie very close to zero. For evolution to occur through mutation, countless sequential good mutations would be required; at each step all would have to cooperate harmoniously and each mutation would have to be selected for. This simply could not happen. As Arthur Koestler has put it:

Each mutation occurring alone would be wiped out before it could be combined with the others. They are all interdependent. The doctrine that their coming together was due to a series of blind coincidences is an affront not only to common sense but to the basic principles of scientific explanation.[1]

In the last century breeders improved the quality of sheep's wool and raised the yield of sugar in beet from 6 to 15 per cent, where it reached a limit. In the present century new varieties of corn, wheat and rice have been developed. Whether or not a person is an evolutionist is unimportant in genetic work. W.E. Lammerts, a breeder who produced prize-winning varieties of rose and worked on camellias, dahlias and lilac, was the chief research geneticist for a large nursery company in California. Yet he was a creationist. Examples of 'evolution in action', such as the peppered moth or Galapagos finch demonstrate variation but not radical, archetypal change. So do successions of ammonites, graptolites and other fossils in the rocks. Indeed, from a mysterious origin graptolites grew *simpler* over time. But they remained graptolites. There is a consistency of information in the genetic book, a point beyond which it cannot be misprinted. In biological terms, there is a limit past which the 'elasticity' of a genome cannot be 'stretched'. Rather than transmute into another type of organism, the mutant either 'snaps' (aborts, is sterile or cannot survive) or 'recoils' back towards the form of the wild-type.

Do we, therefore, ever see mutations going about the business of producing new structures for selection to work on? No nascent organ has ever been observed emerging, though their origin in pre-functional form is basic to evolutionary theory. Some should be visible today, occurring in organisms at various stages up to integration of a functional new system, but we don't see them: there is no sign at all of this kind of radical novelty. Neither observation nor controlled experiment has shown natural selection manipulating mutations so as to produce a new gene, hormone, enzyme system or organ.

Sir Julian Huxley, although allowing a very generous rate for favourable mutation (one in a thousand), calculated that the odds against one million successive, successful mutations which might result (if they were, in no case, bred out or reversed in the meantime) in the evolution of a horse were $10^{300,000}$ to 1. At this rate (and it would take three five-hundred-page volumes to write down the zeros) there is a shortfall by hundreds of orders of magnitude of the estimated age of the earth (less than 10^{18} seconds) compared to the time required for such hypothetical evolution to occur. To argue 'because the horse is here, it must have evolved' is to beg the question.

Teleological theory, which postulates an archetypal blueprint, predicts the intensely conservative nature of the genetic code which, in practice, we find. On these terms, mutation is seen as a negative contributor to genetic diversity, a burden or load upon its 'host'. The catalogue of medical mutational disorders in man, for example, has already passed 1500 and continues to grow. Natural selection may reduce the incidence of genetic disease but cannot eliminate it entirely because most mutations are recessive: some, like the haemophilia gene carried by Queen Victoria, can be paired with a dominant gene that masks the mutant's effect and bypasses natural selection for generations.

It is a wonderful house we live in. The essential structure, materially viewed, is its controlling codes. Where did they originate? Even the bacterium is an architectural triumph. Its bricks are nucleotides, sugars, proteins and other large purpose-built molecules. In eukaryotes these molecules are the product of genetic shuffling (fig. 9) and vary a little from individual to individual, from type to type. In this sense their variation is as much a matter of chance as a poker deal in Las Vegas or a roulette-wheel call in Monte Carlo. Life deals a genetic hand, a physical fate with which each organism has to play the game as best it can. Yet, says the creationist, it is false to argue from this that life evolved by chance. Certainly, an element of chance is built into the game. But poker and roulette are played with exact equipment, according to precise rules, in a clubroom or casino. Similarly, life's game is played with genes and chromosomes, according to mitotic or meiotic rules (fig. 8), in a body. It is the origin of this equipment, rules and house of play which interests the creationist. Did these evolve by chance? Says he, 'No chance!'

The Information Gap

A computer programme or a musical score contains, as opposed to non-sense or cacophony, a great deal of information. Biochemists now realize that life's tissues are invested with a tremendous amount of exact and inter-acting chemical information. It is the origin of this information, which must be precise, that is the problem.

American geneticist S. Ohno claimed that, had evolution been entirely dependent on natural selection, only numerous forms of bacteria would have emerged from a bacterium. The creation of metazoans, vertebrates and, finally, mammals from unicellular organisms would have been quite impossible, for such big leaps in evolution required new blocks of genetic code, with previously non-existent functions, to be 'clipped in' with the old.

Discontinuity is the issue. Where Darwin staked his claim on continuity, discontinuity is a primary prediction of the creationist model; and it cor-relates the known data well. Today we recognize, through the taxonomic system, the distinctions between organisms. For the past, palaeontologists are aware that the fossil record contains precious little in the way of inter-mediate forms. Most new species, genera and families, and nearly all cate-gories above the level of families, appear in the record suddenly and are not led up to by known, gradual, completely continuous transitional sequences. A critical analysis of zoological groups, including fossil and recent species, shows that most of them 'froze' in their present state a very long time ago. In the last 10 to 400 million years, all of them have exhibited only slight variations; no new, broad organizational plan has appeared for several hundred million years.

No one is more aware of the inadequacies of the fossil record than the geologists themselves. As Darwin wrote:

The geological record is extremely imperfect and this fact will to a large extent explain why we do not find interminable varieties, connecting together all the extinct and existing forms of life by the finest graduated steps. He who rejects these views on the nature of the geological record, will rightly reject my whole theory.

Steven Gould, a modern evolutionist, adds a sad postscript:

Palaeontologists have paid an exorbitant price for Darwin's argument. We fancy ourselves as the only true students of life's history, yet to preserve our favoured account of evolution by natural selection we view our data as so bad that we almost never see the very process we profess to study.[2]

Can mutations transmute? The genetic shift from one kind of organism to another requires that a precise chunk of genetic material be plugged in; that various subroutines be tailored into the new programme. Chromosomes are chapters in the book of life, chemicals the paper and ink, and mutations are the bugs in the system. They are like syntax errors, printing errors, smudges and blots on the paper. At best, they represent doubly printed pages or grafts onto an already complete story. But they cannot, in a hundred or a hundred billion years, claim to have written the story. Of whose or what's authorship is nature's book of life?

What Do You See in an Eyeless Fly?

Mutations 'bug' the system. Those which can occur with non-lethal effect in any given species seem limited. This limit proscribes major structural change or nascent organs, which evolution should be producing but are nowhere evident. In fact, as we have seen, both genetic machinery and natural selection act more in the interests of conservation than of creation.

Take the example of fruit flies (*Drosophila*). Morgan, Goldschmidt, Muller and other geneticists have subjected generations of fruit flies to extreme conditions of heat, cold, light, dark and treatment by chemicals and radiation. All sorts of mutations, practically all trivial or positively deleterious, have been produced. Man-made evolution? Not really: few of the geneticists' monsters could have survived outside the bottles they were bred in. In practice mutants die, are sterile or tend to revert to the wild-type.

Arthur Koestler related[3] how one set of induced mutations led to a stock of eyeless flies which, if inbred, within a few generations engendered flies with normal eyes. He could not accept the standard explanation of geneticists that other members of the gene complex have, at random, been 'reshuffled and combined in such a way that they deputize for the missing normal eye-forming gene'. Or, it might be said, deputize for the deficient eye-forming gene-complex. He believed some genetic self-repair system to be in operation, which acts in a way similar to the 'strategy of the genes' whereby an embryo is enabled to reach its goal despite various hazards encountered on the way. As an evolutionist (albeit a rebel who appeared to be attempting to resurrect a limited form of Lamarckism) Koestler interpreted such resilience as 'evolutionary self-repair'. Are the mechanisms of phylogeny, like those of ontogeny or growth, endowed with some kind of evolutionary instruction-booklet?

To the teleologist this is a superfluous question. The case of the fruit fly's eyes is seen as an example of the strong conservative 'elasticity' of the genes to return to their archetype rather than to evolve.

Products of Distinction

Discontinuity shows not only in extant and fossil phenotypes. Organs, cells, organelles, even biochemicals, are distinct in structure and function. For example, the mould *Neurospora*, like others of its kind, manufactures an amino acid, arginine. At least seven stages are involved in the process, each with its own specific enzyme. Some, if not all, of the intermediate stages involve the production of molecules (ornithine and citrulline are examples) believed to be otherwise quite useless to the mould. They are simply means to reach arginine. How did they, their enzymes and precursor genes evolve step-by-step, together, until the end-product, arginine, was achieved? How did natural selection not interfere to eliminate these immediately useless, though stage-wise essential, products?

In a factory different 'shops' use different tools of the trade in a concerted processs which leads to a preconceived end product. Similarly, in a body, cell types do not 'grade' into one another. Distinct cells are biochemically equipped to execute the steps which lead to a biological target – whether it be the production of a hormone like insulin, transmission of a nervous impulse or whatever. The question occurs why intermediate products (such as citrulline), superb machine tools (which enzymes are) or distinct cells should – somehow 'anticipating' the process as a whole – survive the rigours of natural selection in a useless state in order to participate later.

This point is driven home, at visible level, by the intermediate stages in the development of a butterfly. When the feeding activity of a caterpillar ceases, it shrinks and fixes itself by its own silk cords to a stem; many other changes occur before the chrysalis takes its final shape and hardens. Within this dry shell the organs of the caterpillar are dissolved and reduced to pulp. Breathing tubes, muscles and nerves disappear as such; the creature seems to have died. But processes are in operation which remould that pulp into different, coordinating parts, and in due course the insect, which has not grown up or developed in any normal sense, re-emerges as a beautiful, adult butterfly. It is a kind of resurrection. Certainly it demonstrates the absurdity of invoking natural selection by successive mutation to explain such an obviously, yet subtly programmed, process. Why, on that basis, should the ancestral insect have survived the mutations that projected it into the chrysalid stage, from which it could not yet develop into an adult? Where was natural selection then? How could pre-programmed metamorphosis, in insect, amphibian or crustacean, ever have evolved by chance? Indeed, how could development (chapter 8) have evolved piece-meal? The ball is in the evolutionist's court, tangled in a net of inexplicability.

Monsters

An adaptive strategy is required to provide for the discontinuity that Darwin denied but that we, like the creationists, perceive. Such a strategy has been evolved at two levels. At the level of genetics its pioneer was Richard Goldschmidt (1878–1958), an eminent geneticist from Berkeley, California. His major work, *The Material Basis of Evolution* (1940), sparked off controversy which still smoulders on. At the level of macrobiology, Stephen Gould has suggested a theory of 'punctuated equilibrium'. Paradoxically, therefore, evolutionists are driven to seeking to justify a position formerly impugned, although in accordance with the observed facts, as creationist heresy. It must be emphasized that, like abiogenesis, neither hypothesis has been proven by observation or experiment. They serve, however, in seeming to snatch neo-Darwinism from the jaws of creationists.

Creationists agree with Darwin (though not with neo-Darwinians) about the importance of 'sports' (p. 66). Goldschmidt also observed, after forty years working with micro-mutations, that they seemed to lead nowhere. However many mutations are combined in such an organism as the fruit fly (a favourite sacrifice upon the altar of genetics), it remains itself. Goldschmidt was puzzled: if mutations lead nowhere, what are the origins of hair in mammals, feathers in birds and segmentation in arthropods? Where did teeth, blood circulation, compound eyes and the poison apparatus of the snake come from? Logically enough he broke with Darwinian graduation, suggesting that 'mega' or 'macro' mutations must provide the answers. These amounted to such a shake-up in the genetic works as would normally be intolerable – fatal. Goldschmidt argued, however, that once in a while a 'hopeful monster' must appear, burdened with mutation complexes that could, in the course of time, turn into something useful.

Quantum jumps occur in physics where electrons exist at one energy level or another, but never in between. In a wide departure from the compass of modern genetics, never before or after him endorsed, Goldschmidt analogized a quantum jump or 'saltation' in genetic material which might, for example, lead to a prototype bird hatching from a reptilian egg.

The discontinuity implied by a quantum jump was initiated, he said, by a small, underlying genetic change. This occurred in a 'rate gene' which produced its effect by changing the rates of development of the different parts of the embryo. These might be rates of growth or differentiation, rates of production of materials necessary for differentiation or rates of reactions leading to certain physical or chemical environments; if it occurred early in growth, a small change of genetic instruction could result in a considerable change of adult form.

Goldschmidt's suggestion that macro-evolution has occurred as the result of simultaneous, co-beneficial mutations in a few key 'rate genes', operational in the early stages of embryonic development, is difficult to confirm or deny. Has it occurred in the past, resulting in the production of 'new' organisms alongside their parent stock? We do not know because embryos are seldom fossilized, so we cannot see changes in the stages of growth. There is some evidence for neoteny, the precocious development of sexual maturity in juvenile forms, which may give rise to distinctive species. For example, you and I are thought by some to be a neotenous kind of great ape, retaining many 'juvenile' characters in the adult and sexually mature form. But this is speculation. If neoteny and other shifts of growth rate are to explain macro-evolution, they must account for the origins of egg, embryo, organs and the 'perfections' (p. 81) of specialized structures.

Fits and Starts

Controversial Harvard palaeontologist, Stephen J. Gould, believes that the mutations (mentioned above) could make evolution possible. But he differs from population geneticists who, from their mathematical analyses, believe the process happened gradually. Large populations of organisms are normally genetically stable because new and non-lethal mutations are diluted by the sheer bulk of the population through which they must spread. They may build slowly in frequency but changing environments usually cancel their selective value before they have become 'set'. The consequent lack of change over long periods of time is called 'stasis'.

Gould believes, however, that occasional 'punctuations' occur amid long periods of 'stasis'. These 'punctuations' are relatively short periods of time during which inbreeding within small, isolated 'founder' populations, together with rapid environmental changes, stimulates rapid evolutionary change. He updates the previous generation's 'Sewall Wright Effect', a mathematical treatment of the idea that the rules of selection are relaxed in a small population so that unusual things can follow if a few mutations occur. The normal effect would be extinction. Where Goldschmidt's 'hopeful monster' failed, however, the hope is that this 'effect' might generate a new creature without the need for a missing link.

Some geologists believe that the explosive 'punctuations' are associated with geological cataclysms. During the equilibrium, or quiet periods, neo-Darwinian mechanisms may operate to produce the observed minor variations. This is not gradual or smooth, but jerky, step-wise evolution. If change is rapid in small populations, very few transitional forms can expect to be fossilized. Therefore, most attractively, this non-Darwinian theory predicts an absence of link-fossils, which ties in with the facts. It overcomes the problem of fossil gaps.

Gould appears pleased that this concept of revolutionary evolution, which agrees with the Marxist dialectic and Soviet philosophies of change, has allowed us to consider an alternative to western gradualism and unrestrained 'struggle for survival' (in political terms, laissez-faire capitalism). Darwinian evolution was nature's eugenics programme, an idea which upsets Marxist biologists. The dialectical laws of Marxist philosophy are explicity punctuational and Gould believes it is no accident that Russian palaeontologists support a model similar to his.[4]

Naturally, he does not assert the exclusive truth of his philosophy of punctuational change. He cannot because, unlike gradual minor variations, it is not observed. Nonetheless, he believes it may prove to map tempos of biological and geological change more accurately and more often than any of its competitors.

Imperceptible Authorship

If palaeontologist is here at odds with geneticist, there seems to exist another genetic 'bone' of contention. Not all mutations are harmful; some, which have little effect, are called *neutral*. Does the accumulation of mutations, which is believed to occur at a fairly constant rate, contrast with the relatively sudden erratic way that visible forms are supposed, according to Gould's theory, to change?

The book of life is slowly being deciphered. This has led to the realization that there exist three main fractions of DNA. There is unrepetitive DNA, which yields m-RNA for the manufacture of proteins. There is an intermediate type, fairly repetitive, coding for t-RNA and r-RNA, that goes to make ribosomes. Thirdly, there occurs a large amount of very repetitive DNA in simple sequences, seemingly meaningless and 'redundant', although it may be connected with gene regulation or chromosome structure. When, for example, we find a crayfish or salamander with far more DNA than ourselves, we look for possible uses of it; 'gene duplication' is another suggestion. Redundant copies of genes may have accumulated, like duplicate paragraphs miscopied into the book of life. Neutral mutations (which are selected neither for nor against) may alter these 'supplements'. The main programme remains unimpaired so that change in these genetic duplicates is imperceptible until, with a last critical mutation, the new genetic sequence 'clicks' into working order. The relevant new story-line might be a feather, a tooth or a shelled egg.

At least, that is the theory. It is as if you were to complete a *Times* cross-word by picking up and putting down alphabetical counters while blind-fold. In the computer analogy, it is as if, upon the allocation of extra memory, permutations of operands and variables were, without rhyme or reason, added or erased until, at a particular moment, an extensive, fresh and elegant addition to the main programme came into operation.

If the neo-Darwinian wishes to diminish the powerful computer analogy, it is only because it contrasts logically yet sharply with his own belief. It highlights the perennial problem – the origin of life's complex, working programmes. For the neo-Darwinian, the formation of biological, chemical and coded combinations, carrying information, can be explained by the rules of chemistry. But although physical law can explain the operation of a machine designed by man, it cannot explain the origin of the design and the translation of the original blueprint (archetype) into the pro-totype machine. Creative imagination is not reducible to physics and chemistry. A teleologist applies this reasonable argument to natural bio-technology. He sees the DNA molecules as paper and ink, a gramophone or a computer system; and the information (the story-line, music or instructions) as equivalent to the information carried by the DNA mole-cule. He eliminates chance and promotes purpose in the way that intel-ligence, that imperceptible creator, naturally does. More plan, less chance. This gives life meaning.

It is clear that, viewed in this light, 'sports' mean no more to the creationist than they did to Darwin – curiosities, unimportant because they represent disadvantageous modifications from the point of view of a struggle for existence.

The Hone

Why isn't Edward Blyth's name a household word? Why isn't he buried in Westminster Abbey? Blyth (1810–73), a creationist, first published essays on natural selection in 1835, 1836 and 1837, over twenty years before Darwin published *The Origin of Species*. Loren Eiseley found evidence[5] in Darwin's essays that, between 1842 and 1844, he had studied Blyth's work. Later, after Blyth went to Calcutta, Darwin corresponded with him, showing particular interest in his studies of animal variation.

Blyth made no more of his notion of natural selection than the facts war-ranted. He drew attention to its conservative function, using it not to explain how species arose from pre-existing species but rather why they remain constant. To him a type's pedigree was a distinct creation, kept fit and 'in form' against the hone of natural selection. Only the fittest would survive to reproduce. Inbuilt adaptive potential could be exploited – are not negroes dark as excellent protection from the sun, and tubby Eskimoes, with their reserves of fat, considered locally to be the handsome ones? If the strongest lion and the swiftest zebra survived, then this showed not that the species varied, but that they remained constant.

If Darwin did absorb Blyth's ideas, he made no reference to him. Nevertheless, like everyone else in science, Darwin borrowed and elaborated ideas from other people. Evolution came from his grandfather and others of earlier generations; uniformitarianism came from Lyell. Why should he not, with seminal insight, re-express natural selection in evolutionary terms? He turned Blyth on his head, agreeing that natural selection would keep a species healthy but adding that the environment, like a changing mould, gradually changed the species. For Darwin this change was not just a fluctuation around a median but a progressive difference. Selection pressure, the killing power of the environment, 'designs' new species by cutting away unfortunate mutations. The hone grinds down all but the best adapted, clearing the way for not only the strong but the 'innovative' of life's candidates.

Charles Lyell, the geologist to whose ideas and friendship Darwin owed so much, was unhappy with this conclusion. His principle of uniformity held that the fossils, rocks and other features of the earth's crust formed slowly over vast aeons by the same processes, under approximately the same conditions that currently prevail in the earth. So for Lyell natural selection would need to operate within an essentially constant environment. How, in a constant environment, could well-adjusted species have evolved from a common ancestral stock, and be evolving further into new forms even better adapted than themselves?

Darwin was intrigued by what artificial selection by man could do to pigeons and dogs, sheep and cattle; extrapolating, he believed that, in the constant struggle for survival caused by limited resources, varieties which best fitted their environment could, in a similar way, be selected by nature.

But Blyth's idea may well have been closer to the truth than Darwin's development of it, for natural selection can only reduce rather than increase genetic variability. It operates in nature solely as a conservative mechanism, a sieve to weed out the weak, malformed or sick and maintain a healthy stock. It is indeed a force counteracting the tendency for mutation to cause a degeneration in the quality of living organisms – but it cannot be creative.

The Case of the Coloured Birds shows its limitations, and the way it can induce genetic death. Take a population of birds, whose colours closely match their background, living on an island and showing both dark and light plumage phases. There are many examples in nature from red grouse to snow geese. As their population increases, some birds colonize a neighbouring island which is predominantly darker than their original island. The paler sections of the population on this new island are easily picked out by predators and destroyed; the dark birds are less easily seen and survive. Gradually, by this rigorous process of natural selection, a race of all-dark birds develops.

On another nearby island, this time lighter coloured, the paler birds in the population are favoured and a paler race survives. So by natural selection two races of birds have developed from the original mixed population, and eventually these differ enough to be considered new species.

But this is a sorting process, not a creative one. At genetic level, within the original population, alleles or gene combinations for dark, intermediate and pale phases must have coexisted. On either of the new islands this stock of genes became depleted when pale or dark alleles were lost by the death of their owners: thus natural selection, far from creating something new, has simply segregated and impoverished the gene pool. And since each resulting population of birds is genetically poorer, each is more prone to extinction. Environmental changes that altered the colours of an island (new vegetation or weathering might do that) would leave each 'adapted' new species unable to respond further, and vulnerable once again to predators. Operating on a wide scale, this process would, within a shifting environment, lead to the extinction of many species. The geological record shows such extinctions have occurred.

This is Darwinian natural selection at work, but it no more adds up to evolution than it did in 1842, when Darwin could not publish against the strong arguments of Blyth and Lyell. Indeed, it is closer to Blyth's view – natural selection as a sieve to weed out unhealthy individuals and maintain healthy stock, with diversification into species from the initially created organisms a limiting rather than a progressive process.

Nevertheless, Darwinist Sir Julian Huxley could write:

. . . natural selection converts randomness into direction, and blind chance into apparent purpose. It operates with the aid of time to produce improvements in the machinery of living, and in the process generates results of a more than astronomical improbability which could have been achieved in no other way.[6]

Astronomical indeed.

The Survival of 'the Survival of the Fittest'?

In 1798 the Reverend Thomas Malthus argued in his *Essay on the Principle of Population* that competition for limited resources was a law of nature. Such competition would favour healthy individuals and act as a stabilizing force which prevented changes in animals and plants. Darwin turned Malthus, like Blyth, on his head. He argued that such competition, wherein the fittest survived, was the driving force behind natural selection.

Darwin borrowed the phrase 'survival of the fittest' from Herbert Spencer (1820–1903), a Victorian philosopher and polymath who was applying it in studies of sociology, psychology and capitalist economy. Are you muscular? Curvacious? Wily? Fitness is nowadays defined in terms of two components, reproduction and survival, though in practice it has not proved at all easy to measure. In theory, the fittest individual is the one whose offspring survive to reach maturity in the greatest numbers. Why, therefore, haven't herrings and rabbits inherited the earth? Why should bacteria ever have evolved out of their extremely successful but unambitious rut?

No one has any argument with natural selection or the survival of the fittest, as such; but to observe that 'nature selects the fittest' is far from explaining where the fittest come from. It is remarkable that Darwin failed to notice the truth in the converse of what he had said; the catastrophes that end lives – drought, flood, starvation, plague – are non-selective. The strong are struck down with the weak. Is the blackbird's early worm less fit? It has been shown, by night-time photography, that lions do not necessarily seek out the smallest, weakest buffalo. They may take fully adult males. Nor does a predator overkill or extinguish its prey because, if it did, it would itself die. Moreover, if there were no mutant prey, it would catch fit prey. Mutants are either left to die at birth or form such a small proportion as to make natural selection, on these terms, meaningless. This does not in itself invalidate any doctrine, but it highlights the point that all forms that continue to survive (including the apparently weak or vulnerable) have the same survival value.

It is matter of perspective. And the perspective need not be evolutionary. Diversity of species reduces competition. Far from nature 'red in tooth and claw', each creature is skilled at extracting energy in a different way from its own particular niche in the environment; many of them have roles in the ecosystem that avoid competition. As Grassé noted, even in the mud of a pond '. . . cohabitation of species belonging to groups widely different in system teaches us that in one and the same environment separate types of biological system ensure the survival of one and all'.[7]

Certainly, ecological science shows myriad positive mutual relationships; it shows, too, that the disturbance of one species *adversely* affects others in a given region. 'Survival of the fittest' can only lead not to ameliorating but to worsening conditions. A good example is the ecological crisis which surrounds our man-centred technology, where the human survivor presides over ruined landscape, devastated forests and threatened animal species.

One can understand Darwin's interpretation, coming from a time and a place where the cry was 'exploit', and the dominant image was of nature as an enemy against which man battled. Now this has changed; the cry is 'conserve' and the image is of the community of life on earth. In 1855 Chief Seathl of the Swamish tribe wrote to President Franklin Pierce, protesting at the white man's aggressive use of American land.

'What,' he asked, 'is man without the beasts? If all the beasts were gone, man would die from great loneliness of spirit, for whatever happens to the beasts also happens to man. All things are connected. . . .'

No less than Chief Seathl, the creationist sees life on earth as a web, a created whole. The biosphere is an 'organism', comprising many kinds of flora and fauna which interact with the land, sea, air and each other in such a way as to keep the planet a fit place for life. On a lifeless planet you would not find our climate, atmospheric gas content, salinity levels in the sea or any of several other vital 'eco-factors' preserved, without large or biolethal fluctuations, for perhaps several billion years. The maintenance engineers, the fundamental workforce which preserves the balance, are bacteria and plankton. Under the conditions that these micro-organisms sustain, 'higher' life such as our own can flourish; upon this microscopic foundation, whether at once or in stages, the hierarchical biosphere has been built. Although this biosphere buffers the slow 'churning' of the physical environment by generating new species, the framework of its architecture, *types*, does not change. Its wholeness stays.

If the *Beagle* had put into North as well as South America and Darwin had met a redskin like Chief Seathl, he might never have borrowed Spencer's catchphrase, 'survival of the fittest', and popularized it as an evolutionary concept.

Fitting In

It is assumed that an ancestral stock of Galapagos finches reached the islands from the American mainland and then, in the absence of any competition, radiated to form a series of birds occupying quite varied habitats. Their beaks, for example, vary according to their diet. There are ground finches which feed mainly on seeds; others feed on prickly-pear cactus. One species climbs trees like a woodpecker and digs insects from the bark with a stick; another resembles a warbler in diet and behaviour. When several divergent forms, adapted to different modes of life, stem from a common ancestor, it is called adaptive radiation. New features, called adaptations, evolve and permit the invasion of fresh territory or new use of an old environment.

What about these adaptations which natural selection is supposed to produce? Darwin said: 'If it could be demonstrated that any complex organ existed which could not possibly have been formed by numerous, successive, slight modifications, my theory would absolutely break down.'

There is no dearth of candidates, such as flight or sight, but everything depends on the observer's viewpoint. Modern evolutionist Richard Lewontin tackles the problem of the origin of specialized traits.[8] Evolution, he says, cannot be described as a process of adaptation because all organisms are already adapted. In other words the major features of their biological design, such as wings or eyes, have preceded natural selection: natural selection cannot explain their origin.

For the teleologist there is no problem. Adaptation takes two forms, operates at two levels. Adaptive radiation, as exemplified by the Galapagos finches, is a secondary form of adaptation. Such adaptation may simply reflect a changed genome whose product has been able to survive; or, in a changed environment, natural selection may prune products other than those which fit the new conditions. Either way, are these relatively trivial Darwinian adaptations – called by some micro-evolution and by others variation – really the basis of biological design? Or just superficial changes on a deeper, stable archetype?

Primary adaptation is created. It is the deliberate adaptation of an archetype such as the pentadactyl limb (fig. 2): this would have been created differently for roles on land, in water and in air. Or it is the incorporation of genetic or physiological potential to respond to extreme conditions. The ability of moss to dehydrate, survive and then revive at the touch of drops of water is an example.

Lewontin is no teleologist, but he makes a similar point. If a region becomes drier, the plants in it can respond by developing a deeper root system or a thicker cuticle (waxy coating) on the leaves, but *only* if their gene pool contains genetic variation for root length or cuticle thickness. The genes for deep roots and thick waxy coats must be present *before* natural selection can operate on them – just as a potential for colour variation had to be present in the birds we discussed a few pages back. Natural selection enables organisms to maintain their state of adaptation rather than improve it in an 'upward', evolutionary sense: it enables a species to keep up with the constantly changing environment.

In *Alice in Wonderland*, you may remember that the Red Queen had to keep running to stay in the same place. The Red Queen's hypothesis, formulated by evolutionist Leigh van Valen, proposes that, because the environment is constantly changing, the function of natural selection is to enable organisms to 'keep up with it'. As the balance of environmental pressure is, at random, changing, so the fitness of one species relative to another constantly varies. Now one, now another becomes the fittest competitor. The function of natural selection is not so much to 'improve' an organism as to allow it to maintain its state of adaptation. In a new environment the fitness of a previously dominant organism may decrease, even to the point of extinction. In this way van Valen explains what the fossil record shows, that extinctions are random affairs, striking down well-established species and latecomers alike. This fits in well with the creationist view.

A Perfect Fit

'Nothing is wasted' is a phrase sometimes heard in connection with biological systems. Not only is there a correlation between the structure and function of organisms (e.g. whales), organs (e.g. lungs, gut or kidney) and tissues, but even at the cellular and molecular level of life such 'architecture' appears. Indeed, it appears with increased precision, because we can actually measure the items concerned in terms of chemical constitution and enzymic (pp. 144, 258) or other activitiy, and relate structure and function down to almost atomic level.

An example is pheromone sensitivity in moths. Under ideal conditions a male moth can sense female scent at a dilution of only forty molecules per cubic centimetre. It can home in from several kilometres if the wind is in the right direction! Its antennae hairs are sensitive to only one molecule of pheromone. You cannot have less. Sensation is limited by physics rather than biological engineering which is here effectively perfect. Indeed, there is reason to believe that the correlation between structure and function can go even further and that, as in respiratory and photosynthetic systems (chapter 10), nature is exact to the last proton and electron.

Why should a certain complex of atoms, called 'selfish genes', strive after this sort of perfection? Why should DNA weave ever more complex protective cocoons (called bodies) in order to ensure its immortality? What drives a chemical to 'live', reincarnate, survive, improve its chances?

Are organisms like you and me no more than instruments for the reproduction and survival of this remarkable material, no more than temporary
packets in which it posts itself through time? Yet, in the end, survival is the
same as existence, and what is this for?

In an article called 'A Natural Precision Designer' Stephen Gould unin-
tentionally expresses creationism in a nutshell: 'Good design is usually
reflected by correspondence between an organism's form and an
engineer's blueprint.'[9]

In the creationist view, the perfection of that blueprint does not need,
any more than an idea, to have a history. There is no need for evolution by
trial and error. The fittest came first. This is not to say that prototypes
were physically fitter or finer specimens than can exist today; it means that,
from the first, they worked well and were, unlike the coloured birds we
spoke of earlier, rich in created alleles. Their genetic potential was at
maximum.

It is up to the Darwinian to say whether a chemically superior archetype
could have been created. Certainly, perfection of adaptation and structure
has always bothered Darwinians, whose opponents are likely to cite it as
evidence of a Supreme Designer. To suppose that such complex, superbly
adapted organs as the mammalian eye (p. 215ff.) could have been formed
by natural selection seemed, as Darwin freely confessed, absurd in the
highest degree. Kew botanist Alan Radcliffe Smith points out an equally
remarkable case:

The Lady's Slipper Orchid is an example of a 2-stamen orchid. As the common
name implies, the lip is very distinctive, being shaped like a shoe or slipper. The
inside of the lip is very smooth and this, together with the inrolled edges, prevents
the easy departure of an insect visitor by the same way in which it came. Instead, it
is forced by the shape of the lip and by the nature of the surface to move towards
the back, or point of attachment, where there are two small exits. In order to gain
these exits, the insect must first pass beneath a stigma and then brush past one or
other of the two stamens, which deposits pollen onto it, after which it is free to fly
off. If it then goes to another slipper, it will pollinate it with the pollen gained from
the previous one; the second slipper will not be on the same plant as only one
flower is open on a given plant at any one time, and thus cross-fertilization is very
efficiently effected. . . . The complexity of interaction between plant and insect is
truly staggering and, for those who will see, it clearly bears the hallmark of the all-
wise creator.[10]

The challenge in that phrase 'for those who will see' is one that no true
evolutionist would let by: however 'perfect' in function, orchids could just
as well have evolved from other flowers. Stephen Gould comments:

If God had designed a beautiful machine to reflect his wisdom and power, surely he would not have used a collection of parts generally fashioned for other purposes. Orchids were not made by an ideal engineer; they are jury-rigged from a limited set of available components.[11]

It is the kind of argument we have heard before, and the creationist weighs in on cue. They could equally have been created according to thematic design. It is untrue that inventors eschew parts fashioned for other purposes to produce their novelties, as the records of the Patents Office would show. Only rarely is there some miracle of new design − a wheel, for example, or a zip fastener; normally, ingenious adaptation is the order of the originator's day, and a Creator might well choose to create economically. Indeed, any ordering influence behind the programming of life's codes (genotypes) and consequent forms (phenotypes) might be expected to practise economy in proportion to its intelligence. Efficient, non-chaotic short-cuts would be used in all possible syntheses, not least in coding. It would, for example, be logical to use one of a few basic plans of construction for molecular biology, physiology and anatomy, and then to vary them. In this way the creationist hopes to indicate that bodies are not just 'gene machines'. Genes are as important as a score-sheet is to a musician, but it is by the fact of its purposeful creation that life has real meaning.

A Fitting End

Alfred Russel Wallace (1823−1913) started it. He sent a paper from the East Indies to Darwin called 'On the Tendency of Varieties to Depart Indefinitely from the Original Type'. It arrived in June 1858; so clearly did it expound 'his' theory that it shocked the recipient. With Hooker and Lyell, Darwin arranged for it to be read at the meeting of the Linnaean Society on 1 July 1858 − after a reading of extracts from his own writings. Immediately afterwards Darwin organized these into *The Origin of Species*, published in 1859.

But Wallace, co-founder of modern evolutionary theory, disagreed with Darwin over natural selection. For example, he observed that supposedly backward natives had languages more complicated than those of modern Europe − a process of linguistic devolution seemed to have occurred. The natives had mental powers far in excess of what they needed to carry on the simple food-gathering techniques by which they survived. 'Natural selection', he wrote, 'could only have endowed the savage with a brain a little superior to that of the ape, whereas he actually possesses one very little inferior to that of the average member of our learned societies.'[12]

There are many instances of children straight from Stone-Age societies, such as the Aborigines, gaining degrees and diplomas. By insisting that artistic, mathematical and musical abilities could not be explained on the basis of natural selection, Wallace challenged the whole Darwinian position. Why, if man's brain was its present size fifty thousand years ago, did it take so long for him to 'warm up' to the intellectual revolution of the past five thousand years? Indeed, why should such an instrument as the human brain have been developed in advance of the needs of its possessor? For a creationist this would provide evidence of pre-programming. For Darwin it was dangerous and he wrote to Wallace: 'I hope you have not murdered too completely your own and my child.'[13]

This chapter has criticized the neo-Darwinian mechanisms of evolution, viz. mutation and natural selection. Let Grassé provide the footnote.

How can one confidently assert that one mechanism rather than another was at the origin of the creation of the plans of organization, if one relies entirely upon imagination to find a solution? Our ignorance is so great that we dare not even assign with any accuracy an ancestral stock to the phyla Protozoa, Arthropoda, Mollusca and Vertebrata. . . . From the almost total absence of fossil evidence relative to the origins of the phyla, it follows that an explanation of the mechanism in the creative evolution of the fundamental plans is heavily burdened with hypotheses. This should appear as an epigraph to every book on evolution.[14]

Notes

1 Koestler, A., *The Ghost in the Machine*, Picador, 1975, p. 129.
2 Gould, S.J., *The Panda's Thumb*, Norton, 1982, pp. 181–2.
3 Koestler, A., op. cit., pp. 133–4.
4 Gould, S.J., and Eldredge, N., *Paleobiology*, vol. 3, Spring 1977, p. 145.
5 Eiseley, L., 'Charles Darwin, Edward Blyth and the Theory of Natural Selection', Proceedings of the American Philosophical Society, February 1959, pp. 94–158.
6 Huxley, J., *Evolution in Action*, Harper and Bros, New York, 1953, pp. 54–5.
7 Grassé, P-P., *The Evolution of Living Organisms*, Academic Press, 1977, p. 178.
8 Lewontin, R.C., 'Adaptation', *Scientific American*, vol. 239, no. 3, September 1978, p. 159.
9 Gould, S.J., 'A Natural Precision Designer', *New Scientist*, vol. 84, no. 1180, November 1979, p. 446.

10 Radcliffe Smith, A., 'Orchids', *Creation Science Movement Bulletin*, no. 216, 1977, pp. 1–2.
11 Gould, S.J., *The Panda's Thumb*, Norton, 1982, p. 20.
12 Wallace, A.R., quoted by Norman Macbeth in 'The Question: Darwin Revisited', *Yale Review*, June 1967, p. 629.
13 Ibid.
14 Grassé, P-P., op. cit., pp. 17, 31.

6 Monkey Business

Let us take a break from cells, chromosomes and genes and look at a different side of the story — the side that Darwin initially found hardest to think about — the origins of man himself. 'On which side — grandfather's or grandmother's — were ape ancestors to be found?' asked Bishop Wilberforce in Oxford. 'On every side,' answered the Darwinists. 'We'll comb the fossil record and find them.' What were they looking for? The break between modern man and modern apes is complete, but at some time in the past must there not have been ancestors that were half ape and half man? The search for a 'missing link' was on. Distinguished amateurs deceived the professionals. Distinguished professionals deceived themselves — even a Jesuit priest was caught up in the monkey business surrounding the search.

The Rising Son?

It is usual for a crucifix, raised in a conspicuous position, to beam its message to the villagers of France. But beneath an overhanging cliff, above the inhabitants of Les-Eyzies-de-Tayac in the Dordogne, there looms another sign. Across from the National Museum of Prehistory the statue of a brute Neanderthal silently 'preaches' evolution.

Joachim Neander (1650–80), foremost hymn-writer of the German Reformed Church, unwittingly gave his name to Neanderthal man. Neanderthal is a limestone gorge, near the village of Hochdal between Dusseldorf and Elberfeld, through which flows the Dussel river. Neander's love of nature used to lead him to this ravine where, it is said, he composed many of his hymns. Almost two centuries later workmen, quarrying in the valley, uncovered the skeleton of so-called Neanderthal man.

It was first brought to the notice of a scientific body in 1857 by Professor D. Schaafhausen who concluded that, despite some interesting characteristics, Neanderthal must be considered human and normal. In 1864 the

German anatomist Mayer pointed out that pathology of the left arm indicated that Neanderthal had been afflicted with rickets, caused by lack of vitamin D. This in turn had caused the eyebrows to pucker causing browridges. In 1872 Rudolf Virchow, father of modern pathology and anthropology, presented a closely reasoned paper demonstrating that the skull and limb-bones were not ancient at all, but from a man who had suffered from rickets (nearly every Neanderthal child examined has shown rickets), arthritis in old age, and several great blows on the head.

Since that time similar bones have been found in China, Central and North Africa, Iraq, Czechoslovakia, Hungary, Greece and northwestern Europe. Over a hundred different bodies have yielded fragments. It seems that Neanderthals were a race of men who suffered from malnutrition. They had prominent eye ridges, a low forehead and a long narrow brain case. Skulls found more southerly are less Neanderthal in character than those found in the colder north.

In 1908 a Neanderthal burial was found at Le Moustier in southeast France. The classic remains were exhumed near La Chapelle-aux-Saintes and studied by Marcellin Boule from the National Museum of Natural History in Paris. Between 1908 and 1913 Boule (under whom Teilhard de Chardin studied) issued a series of scholarly papers on Neanderthal man, climaxed by a massive monograph in three parts. Boule's ape-like reconstruction had enormous influence and represented a triumph for evolution over those who believed the La Chapelle-aux-Saintes remains were of a diseased human.

Elliott Smith, anthropologist at University College, London, joined in. He wrote about 'uncouth and repellent' Neanderthal man whose nose was not sharply separated from the face, but was more like a snout. Stereotypes portrayed (and portray) a shuffling, shaggy hunchback with stupid gaze. It is now recognized that no evidence whatsoever exists to substantiate this fabrication and the reader must judge, in such instances, where science leaves off and humbug begins.

Normal human brain size is 1450–1500 ccs; Neanderthal's is 1600 ccs. If his brow is low, his brain is larger than modern man's. Learning this, Boule resorted to the Victorian pseudoscience of 'phrenology', a theory that the mental faculties are shown on the surface of the skull. He convinced himself that the frontal lobes were inferior in organization to those of modern man. But Boule made mistakes. These were not discovered until 1957 when two anatomists, W. Strauss and A. Cave from St Bartholomew's Hospital medical college in London, took a second, closer look at the fossil from La Chapelle-aux-Saintes. Although the fossil was supposed to be typical Neanderthal, Strauss and Cave discovered that this particular person had suffered severe arthritis, affecting the structure of vertebrae and jaw. Boule should have detected this. The foot was not 'a prehensile organ', the neck vertebrae did not resemble those of a chimpanzee, nor was the pelvis ape-like. Boule had arranged the foot-bones so that the big toe diverged from the other toes like an opposable thumb. This was the cause of the belief that Neanderthal had to walk on the outer part of his foot like an ape. Boule's interpretation of the knee-joint, resulting in the 'bent-knee' gait, was also incorrect. Strauss and Cave punctured the myth he had created. And Virchow's diagnosis was vindicated – the bones from La Chapelle-aux-Saintes were from an osteo-arthritic geriatric with a skull of capacity 1600 ccs.

The Arunta tribe of Australian Aborigines in fact possess, as noted by Thomas Huxley, a large 'Neanderthaloid' skull and teeth, and some grow extra molars. In late 1972 about thirty skeletons were reported from Kow swamp site in Australia. Dated at about 10,000 years, they were of *Homo erectus*. If this is true, *Homo erectus* and modern man were contemporaries. But is it not possible they were Aborigines? For example, although western-ers' teeth are smaller, Aborigines show little reduction in size from *Homo erectus*. Today, in a generation, such 'savages' take degrees at university and practise science. They are men and women, no less intelligent than ourselves.

Culture? How can you tell? Neanderthal burial took place and certain artefacts associated with the remains suggest the craft of tannery. Trepan-ning? A Neanderthal skull was found buried deep inside a cave in Italy with a neat round hole bored into it. Bolas, such as may have been used in hunting game, have been found. Hunting magic may have centred in a bear cult. At the Drachenloch cave in Austria, at an altitude of 8000 feet, a cubical chest of stones covered with a large stone slab was found. Inside were seven bear-skulls all with muzzles facing the cave entrance while six more were mounted in niches on the wall. In another cave in Bavaria the skulls of thirty persons, with ornaments of deer-teeth and shells, had been placed in the earth. The charred remains nearby seemed to indicate cre-mation. The sun of life sets in the west and there, in later Arthurian legend, lay Avalon and the isles of the blest. Whatever the symbolism, the Nean-derthal skulls were sprinkled with red ochre and all faced west.

In the 1960s a Neanderthal man was found buried on a bier of hyacinths, hollyhocks and other species identified from the pollen on the flowers and known for their medicinal properties. A ceremony held, apparently 60,000 years ago in the Shanidar cave on the Iraqi side of the Zagros Mountains, seemed to clinch the matter. Neanderthal was a cultured man.

Years before this the Victoria Institute had noted that, in 1930, the skull and skeleton of a criminal executed in 1892 were exhumed in Australia.[1] The bones exhibited anthropoid ape characteristics, yet it was a fully modern man. By 1975 Neanderthal had been upgraded to a human type.

Neanderthals were no more different from modern man than the various races of modern man from each other. The famous Cro-Magnon artists who painted at Lascaux and elsewhere in France were truly modern. We must have ancestors, and Cro-Magnon is certainly one of them. He inhabited the grassy plains of Europe, North Africa and Asia from, by conventional dating, 50,000−8000 BC and in no way differs from us. In fact, finds in caves on Mt Carmel (Israel), at Magharet-et-Tabun and Mugharet-es-Skuhl, show a mixture of Neanderthal and Cro-Magnon types, strongly suggesting an interbreeding.

Pithecanthropus

The search for a missing link began in earnest when Haeckel − the German philosopher and writer we met in chapter 4, entered the scene. Haeckel hypothesized an imaginary ape-man and blessed it with the name *Pithecanthropus alalus* − the speechless ape-man. He thought it might be found in Southern Asia, or maybe Africa, and even commissioned an artist to paint a picture of this dumb Caliban and its baleful mate. By 1887 his zeal had fired a pupil to sail for the East Indies. Eugene Dubois was keen, even determined, to hunt for and find this long-awaited missing-link.

In Sumatra Dubois found little. But, hearing that a fossil skull had been found in neighbouring Java, he sailed over and bought it; then found another fragment on its site of discovery at Wadjak. The brain volumes of men and women can vary from 790 to 2000 ccs. Apes vary from about 90−685 ccs, according to body size. The Wadjak skulls were fossilized but, being 1550−1650 ccs in capacity, were too like modern man to be claimed as missing links. Dubois bided his time. In 1890 at Kedung Brabus he found a jaw containing the root of a tooth. In September 1891 at Trinil his Malay coolies unearthed a large molar tooth and a month later the fossil skull-cap of an ape-like creature − now famous − was discovered. In August of the following year he found a human femur (also completely fossilized) about fifty feet away, and another molar about ten feet from where the skull-cap had been found; several more human femurs were also dug up.

After correspondence with Haeckel, he ignored the human skulls from Wadjak and the human femurs, except one. This, he decided, belonged to the owner of the skull-cap; he declared them both to belong to a creature which seemed admirably suited to the role of the missing link. Haeckel, without seeing the evidence, immediately telegraphed back: 'From the inventor of *Pithecanthropus* to his happy discoverer.'

Von Koenigswald, who excavated in Java from 1931–40, wrote: 'Dubois' find came at just the right moment at a time when the conflict around Darwinism was at its height. For the scientific world it constituted the first concrete proof that man is subject not only to biological but also to palaeontological laws.'[2]

So, in confusion and secrecy, the myth of Java man was propagated – invented at just the right time to clinch the evolutionist case. Von Koenigswald, in *Meeting Prehistoric Man*, also notes: 'When Dubois issued his first description of the fossil Javanese fauna he designated it Pleistocene. But no sooner had he discovered *Pithecanthropus* than the fauna had suddenly become Tertiary.' Dubois wanted his 'man' to be early – as close to the true ancestral stock as could be managed. So, continues Von Koenigswald, 'he did everything in his power to diminish the Pleistocene character of the fauna . . . The criterion was no longer to be the fauna as a whole but only his *Pithecanthropus*,'[3]

In 1894, it should be mentioned, Dubois published a paper about *Pithecanthropus erectus* which did not mention the Wadjak skulls. That would have cast doubts on the ape-like Trinil skull-cap, whose capacity he had estimated as between man and ape, at 850–900 ccs! When, in 1895, he exhibited the skull-cap and thigh-bone in Berlin, however, Virchow (who had no time for his former pupil, Haeckel) refused to chair the meeting, saying that in his opinion the bones represented a giant gibbon, not early man at all. The thigh-bone, he said, had not the slightest connection with the skull.

In 1921 Australian anthropologist, Professor Smith of Talgai, Queensland, claimed to have found the first Australasian. This provoked Dubois to show the Wadjak skulls; and in 1924 he gave details of the jaw found at Kedung Brabus in 1890, which in 1891 he had called human but now identified as part of *Pithecanthropus*. Only in the early 1930s were the other human femurs announced. The earlier suppression of human material was calculated to highlight the dubious 'ape-man' bones. The irony is that, although by 1936 Dubois himself said he thought that *Pithecanthropus* was a large gibbon, photographs of painted models of slouching 'Java man' were being reproduced in children's and higher level textbooks. They still are, despite the criticism of Boule and Vallois that such models are 'pure flights of fancy'. In the popular and scientific press *Pithecanthropus*, beginning as a dream of Haeckel, engendered its own growth industry and now continues to thrive on it.

Skullduggery

If the *Pithecanthropus* story was unsavoury, that of the Piltdown forgery is an intrigue worthy of a 'whodunnit?' – a tangled web of spurious evidence, innuendo and suspicion. It will surely rank as one of the most notorious scientific frauds of all time. Three recent, reasonably brief detective accounts each 'point the finger' slightly differently.[4] Where this book can no more than summarize the main events, they are well worth a closer look.

Charles Dawson, a Sussex lawyer, was well known locally as a collector of fossils. In the 1880s he presented a collection of fossil reptile bones to the British Museum and struck up a lifelong friendship with the austere keeper – Arthur Smith Woodward. This self-made man, unpopular at the museum due to his aloofness, was an expert on fossil fish bones. He was certainly not an anthropologist; there were few people in Britain at this time who could fill that role, and he was not one of them.

Edwardian pride took a knock in 1907 when a very large jaw with small, modern-looking teeth was found in the Gunz-Mindel interglacial deposits of Germany. Heidelberg man, as it was called, was identified as the oldest fossil remains of man so far discovered. The search for similar fossils elsewhere in Europe was intensified.

In the following year Charles Dawson claimed that he had unearthed several fragments of thick fossilized skull, stained with iron for their full thickness, from a gravel pit at Piltdown, about halfway between Uckfield and Haywards Heath. In 1911 he claimed to have found several more skull fragments (which matched the first), together with other fossils and artefacts from the site. Early in 1912 he informed Smith Woodward at the Museum that he could equal Heidelberg man. In consequence, from May 1912 these two went several times to Piltdown, accompanied from time to time by other members of the British Museum staff. Another who occasionally worked with them was a young French geologist, Pierre Teilhard de Chardin (1881–1955), a Jesuit priest with a passionate affinity for fossils and a renegade interest in evolution – including the evolution of man.

Teilhard's childhood had been spent in the Auvergne mountains of France. From his mother, who was a great-niece of Voltaire, he learnt spiritual intensity; from his surroundings, love of the 'permanent' rocks. When the Jesuit Order was expelled from France to Jersey he was, as a novitiate, so entranced by the fossils he found that he began to doubt his priestly vocation. His fears that palaeontology might conflict with Christianity seem to have been assuaged and he was sent from 1906–8 to lecture in physics at Cairo University. Here he is almost certain to have heard of any interesting sites in North Africa. It is reported that he stayed at Ichkeul near Bizerta in North Tunisia, a site where *Stegodon* (elephant) fossils are plentiful.[5]

In 1908 he came to Ore Place seminary in Hastings, where he was ordained. Again the idea of evolution, which became a master sentiment, began to work in him and he began to search for fossils. In one of the local quarries, on 31 May 1909, he met and befriended Charles Dawson.

Now, three years later, Dawson, Smith Woodward and Teilhard unearthed more skull fragments and other fossils. One day Dawson, in the absence of Teilhard but before Smith Woodward's very eyes, hit gold in the form of an ape-like jaw-bone. It was unfortunately broken in two places, at the hinges and on the point of the jaw, which made it difficult to establish any clear relationship with the rest of the skull that Dawson was assembling from fragments. The canine teeth, which would have helped to identify it as ape or man, were also, unfortunately, missing.

The finds were announced before the Geological Society in London on 16 December 1912. *Eoanthropus dawsonii* (Dawson's Dawn man) was reported as the British answer to the German jaw, and toasted in a thousand sophisticated drawing-rooms. Was not the first man British? Not only Britain's kudos, but its discoverer's too, was considerably enhanced.

On 29 August 1913 Teilhard stayed overnight with Dawson and went next day with him and Woodward to the Piltdown pit. Lo! There appeared one of the two missing canine teeth. Arthur Smith Woodward reported that they excavated a deep trench in which Father Teilhard was especially energetic. When he exclaimed that he had picked up a canine tooth, the others were incredulous, telling him that they had already seen bits of iron-. stone that looked like teeth on the spot where he stood, but Teilhard insisted that he was not deceived. They left their digging to verify his discovery; there could be no doubt about it – Teilhard had found a canine from the previously discovered jaw.

The tooth was pointed like an ape's but worn in ways that suggested human origin. It closely resembled the canine in an imaginative reconstruction that Smith Woodward had caused to be created for the 1912 Geological Society meeting. It was the decisive missing clue and was reported, along with other animal fossils, including a third piece of stegodon tooth, at the next (1913) meeting. Interestingly, the stegodon tooth was found to contain 0.1 per cent uranium oxide and to have a particularly high level of radioactivity, unusual for European fossils but found in those from Ichkeul, Tunisia. The Ichkeul site was not publicly identified until 1918, so that it is very unlikely the Englishmen had visited it. But Teilhard had.

Not everyone was impressed by the Piltdown finds. There existed considerable suspicion in the minds of two amateur palaeontologists from Sussex, Captain St Barbe and Major Marriott. They had, on separate occasions, surprised Dawson in his office staining bones, and did not trust him. On 13 November came an independent opinion that doubted the find. *Nature* published a letter from David Waterston of King's College, London, which ended: 'It seems to me to be as inconsequent to refer the mandible and the cranium to the same individual as it would be to articulate a chimpanzee foot with the bones of an essentially human thigh and leg.'

But in 1915 a postcard from Dawson to Smith Woodward proclaimed more fossils from a second Piltdown site. This site is not identifiable today. The cranial bones were more fragments from the skull that was found in the first pit. Then in 1916 Dawson died, and in the following year Smith Woodward published the Piltdown II finds.

The British Museum authorities, in a manner not designed to allay suspicions, kept the Piltdown fossils under lock and key. Even such anthropologists as Louis Leakey were only allowed to handle the casts (from which any fraud could not be detected).

Only in 1953, when a method for determining the ages of fossil bones by analysing their fluorine content was revived, were critical tests carried out on the Piltdown fossils. The human skull and ape-like jaw, contradicting updated ideas about man's evolution, had aroused suspicion. They were found to be of commensurate age but much younger than the other fossils and artefacts with which they were supposed to have been found. Most specimens had been artificially stained with bichromate. The canine tooth had been filed, coloured and packed with grains of sand. An elephant bone associated with the skull, of a type found also in the Dordogne and Egypt, had been cut with steel instruments into an improbable 'bat' shape. The Piltdown 'skull' was a deliberate fraud.

The faking of the mandible and canine was skilful and deliberate: it was certainly a determined and unscrupulous hoax. Sir Solly Zuckerman, who later reviewed the story, considered the hoaxer knew more about primate anatomy than the experts who were several times deluded. When in 1953 Dr Kenneth Oakley, in collaboration with J.S. Weiner and W. le Gros Clark, completed their investigations and unmasked the fraud, concern was such that a motion was tabled in the House of Commons 'that the House has no confidence in the Trustees of the British Museum . . . because of the tardiness of their discovery that the skull of the Piltdown man is a partial fake'. After all, the Nature Conservancy had (in 1950) just spent a lot of taxpayers' money tidying up the Piltdown site, which had then been declared a National Monument.

Whodunnit? Dawson was certainly at the bottom of it but probably didn't know enough of primate anatomy to do the job well. Others from the British Museum may have been involved, and Teilhard de Chardin almost certainly fixed and planted the tooth. For further clues start with the authors recommended on p. 100, Note 4. Piltdown is a case, if ever there was one, which illustrates the imposition, by 'scientific' evolutionists, of strong hope, desire and prejudice on a few bones.

An Elephant and 'Nellie'

Teilhard's desire to promote evolution, by whatever dubious means, did not end with Piltdown man. Let us return to Java, to Trinil, to the site of Dubois's gibbon skull. In 1907–8 Professor Lenore Selenka led an expedition which, engaging convict coolies, removed 1000 cubic metres of earth from the exact location of Dubois's discovery. Although forty-three boxes of fossils were despatched to Europe, no confirmatory trace of *Pithecanthropus* (Java man) was found.

In 1936–39 G.H.R. von Koenigswald explored the Sangiran area with the help of natives who were offered a reward for every piece they found. As well as this temptation to poor natives to plant bones, inadequate supervision, due to his station at Bandung being over 200 miles from Trinil, left further scope for error. He found fragments of jaw-bones, teeth, skulls and a skull-cap but no limb-bones; these pieces were designated *Pithecanthropus* II, III, IV. Boule and Vallois both report on the ape-like nature of these fragments. Others, such as le Gros Clark, were very critical of the considerable confusion which surrounded their naming, reconstruction and classification.

Teilhard had been 'banished' on a mission to China by the Jesuits, not for fraud (which remained unsuspected) but because of his success as an evolutionary scientist. From 1923–46, as a consultant to the National Geographical Survey in China, he travelled the land extensively. But he was based at Tientsin with Father Licent in the Geological and Botanical Museum. In 1935, at von Koenigswald's request, he arrived in Java hoping to establish a palaeontological connection between Java and China. Although up to that time no fossil links had been discovered, it seems that he and von Koenigswald entered a cave in the Patijan area and found what proof they needed on the floor. What luck! An abundance of isolated teeth – orang, large gibbon, bear and so forth – were 'absurdly' similar to the fossil-bearing deposits of Kwangsi. For the first time orang, gibbon and bear were found in Java. A correlation was established, it seemed, with southern China.

It was hoped the teeth on the floor of the cave would be seen to link Java with Peking man (see below), already discovered 3000 miles away. Teilhard also hoped that the discovery of *Stegodon* (elephant) fossils in the Karst lands of Patijan would establish a connection between the Solo basin (Java) and the Kwangsi fissures. Shortly afterwards von Koenigswald found just what he needed. We are ignorant whether the fossil's radio-activity was commensurate with that of Ichkeul fossils. Nevertheless, another plank was added to Teilhard's 'theory of the origin of man'.

Peking man? In 1903 Professor Schlosser of Germany had, on examining a number of fossils purchased from a druggist's shop in China, found a tooth he considered anthropoid and suggested that early man might be found on that continent. The tooth has since disappeared but it may have served to spark interest. It seems that, as early as 1912, Father Licent had directed attention towards Choukoutien (Dragon Bone Hill, where the Chinese for fossil is 'dragon bone'), thirty-seven miles from Peking. In 1922 two molar teeth were found. From 1929, however, the Rockefeller Foundation granted $20,000 per year to carry out excavations. Dr Davidson Black (who had been an eager visitor to Piltdown) represented the Foundation. Teilhard was an unofficial observer.

On the basis of the 1922 teeth, Davidson Black made extravagant claims about 'primitive man'. In 1927 another molar tooth was found and, on this basis alone, Davidson Black announced '*Sinanthropus pekinensis*' or Peking man. This marks the start of a complex, unsatisfactory affair upon which, as in the Piltdown case, the last word has not been said. Boule wrote: 'Black, who had felt justified in forging the term *Sinanthropus* to designate one tooth, was naturally concerned to legitimize this creation when he had to describe a skull-cap.'[6]

In 1928, 575 boxes of bones were sent from the limestone hill to Peking. In 1929 the Rockerfeller Foundation agreed to finance a Cenozoic Labo-ratory with Davidson Black as Director and Teilhard as advisor and colla-borator with special reference to geological and dating concerns. In all, about thirty skull-bones, eleven jaws and 147 teeth of the 'ape-man' were recovered. In only five cases were sufficient bones of the cranium available to allow a skull to be reconstructed adequately for measurement which, each time, turned out around the 'correct' 1000 cc mark. The fragments were found in an enormous (seven-metre high) heap of ashes. They were mixed up with animal remains, mostly deer and other edibles. There was no difference between bones found at the top and bottom of the deposit, i.e., no evolution was apparent.

In 1929 Mr W.C. Pei, then in charge of excavations, found a skull which was chosen to represent *Sinanthropus*. It was not until 1931, however, after the visit of his former tutor Elliot Smith, that Davidson Black issued the model and a long article for which he was duly elected a Fellow of the Royal Society. His official description in fact differed from an eager, prior despatch by Teilhard, published in the July 1930 issue of the French *Revue des Questions Scientifiques*. In this despatch he notes small brain capacity and apish appearance. The face below the eye-sockets is missing. In Davidson Black's model, though, only the skull-cap, from which it was not possible to check the alleged brain capacity, was present. The skull-cap closely resembled Dubois's Java man (a gibbon). This was the reconstruction on which Davidson Black had worked long, secretively and hard, by night.

A first-hand report by Abbé Breuil of his 1931 visit to the site demonstrated that there had existed at Choukoutien an industry of a nature far too large and advanced for it to be attributed to the small-skulled animals called *Sinanthropus*. Thousands of chipped quartz stones had been transported several miles and used as tools. Enormous furnaces had been kept burning for long periods, leaving the large deposits of ash. Bones had been worked and cut to a level comparable with that of Neanderthal man. Breuil also mentioned 'bolas' stones, used for entwining the legs of creatures, which it would have taken considerable ingenuity to conceive and then construct.

Boule was also invited to the site but when he saw that the only evidence provided was battered monkeys' skulls, each with a hole in the top, he was vexed, denounced Teilhard and ridiculed the idea that the owners of the skulls could have carried out the large-scale industry revealed. He wrote: 'We may therefore ask ourselves whether or not it is overbold to consider *Sinanthropus* the monarch of Choukoutien when he appears in its deposit only in the guise of a mere hunter's prey, on a par with the animals by which he is accompanied.'[7]

Another Jesuit who was in China at the time, Rev. Patrick O' Connell, also denounced Peking man. Choukoutien was, he believed, quarried; lime-burning was carried out there a few thousand years ago. In the course of time the hill was undermined and a landslide occurred which covered everything with thousands of tons of debris. With the aid of the Rockefeller grant this debris was removed and the remains found. Fossil skeletons of baboons and macaque monkeys, which do not differ from contemporary forms except for their greater size, had been found in the district. He believed that these large monkeys were captured for food.

Their meat would be too tough but the heads would be brought back for the brains which, when cooked, would be a delicacy. This would explain the holes in the skulls, the fact that only *Sinanthropus* skulls and jaws have been discovered, and their situation in the ash after being thrown on the fire with other culinary garbage.

In 1933 Mr Pei did discover skeletal remains (with three skulls) of six humans in an 'upper cave', an S-shaped vertical recess probably formed when the landslide occurred. On 15 March 1934 Dr Davidson Black, on entering his laboratory in Peking to examine these fossils, fell down dead among them. His successor, Dr Franz Weidenreich, in charge from 1934–40, denied that there were any natural caves in the hill. He did not mention the bones until 1939.

Dr Weidenreich corrected Davidson Black's figures for the first *Sinanthropus* skull, referring to the specimen as a cast. He also rejected Davidson Black's cast of the mandible, pointing out that it was made of parts of two different mandibles (one of an adult, the other of a young specimen) in order, as he said, to make it look human. Then he set about a second model of *Sinanthropus*. Although Teilhard had already mentioned Pei's 1933 find in *Revue des Questions Scientifiques*, in a later article in the French Jesuit periodical, *Etudes*, dated July 1937, he says Mr Pei had found 'three complete skulls of *Sinanthropus* and portions of others'. These were presumably the humans, and it was from the largest of these (the great male, 1200 ccs) that Weidenreich made his cast.

Or did he? No photographs of the skulls found in 1936 have been published and there is now no trace of them. Did he instead use skull XI, which was not complete and consisted of a number of fragments, to reconstruct a complete, imaginary skull and mandible? From whichever of the

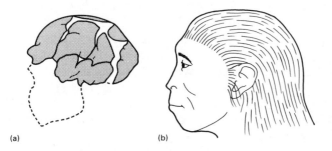

(a)   (b)

Figure 10. (a) *Shull XI*, a reconstruction from broken fragments.
(b) '*Nellie*', a sculpture based on skull XI and other fragments, including a lower jaw found 25 metres higher up

two it was, a sculptress resident in Peking at that time, Mrs Lucille Swann, created 'Nellie' (see fig. 10). Weidenreich commented on the model, 'the most striking peculiarity is . . . the thickness of the neck'. There was no evidence whatsoever for this imaginatively brutish characteristic, yet it is Nellie's profile which most often illustrates *Homo erectus* in school and college textbooks as part of the anthropoid sequence supposed to lead to *Homo sapiens.*

Teilhard all along championed the idea of Peking man. He consistently suppressed mention of the large ash deposit, pointed out by Abbé Breuil. In 1942 he lectured Chinese students on how they had evolved from animals and in 1943 had the lecture printed in pamphlet form. He was all the while in contact with Davidson Black (whom he knew from Piltdown days) and Weidenreich.

The whole collection of fossils disappeared during the war. Some say they were destroyed in transit to America at the time of Pearl Harbor in 1941, others that they were held by Tokyo University at the end of the war. The circumstance of their loss is shrouded in mystery. Teilhard never appears to have given his account of the events and the authorities seem uninterested. Mr Pei, working under Communist rule, published an article in 1954 which described the three rooms in which articles from Choukoutien were displayed. In the first were the casts or models of a few of the skulls made by Davidson Black and Weidenreich. In the next were fossil remains of various animals. In the third, stone instruments. Thus everything except the crucial fossil remains of *Sinanthropus* have been preserved.

Mr Pei said the fossils were shipped to America but Rev. O'Connell believed he destroyed the evidence after Weidenreich left in 1941 and before the Chinese Government returned to Peking. This was to avoid an accusation of fraud because the models did not correspond with descriptions of fossils, published by Boule and Abbé Breuil. Mr Pei did, in fact, resume his work of excavation under the Communists and in 1966 a skull, with an extremely large gap between front and rear pieces, was hailed by the Chinese authorities as that of Peking man.

O'Connell's interpretation of the facts was that several thousand years ago a large-scale industry of quarrying limestone and burning lime was carried on at Choukoutien. It was carried on at two levels. The lime was burnt (as it still is where coal is lacking in China) by grass, straw and reeds, leading to great quantities of ash. Thousands of quartz stones were brought from a distance to construct lime-kilns; they were found at both levels with a layer of soot on one side. Such large-scale lime-burning, it was presumed, serviced the needs of the ancient city of Cambulac on the site of the present city of Peking.

One loose thread. On 16–17 December 1929 reports appeared in the *Daily Telegraph* and *New York Times* of ten skeletons that had been found at Choukoutien. *Nature* announced the discovery on 28 December, saying that Davidson Black would make an important statement on the 29th. No more was heard from anyone on the subject. Was it a mistake? Were the skeletons found to be human? Why have the reports not been queried? Surely Teilhard or Davidson Black could have given us a clue? It may at least be regarded as certain that *Sinanthropus* was derived from gibbon and macaque skulls.

Haroldcookii and the Scopes Monkey Trial

An idea in the ascendant sweeps man with it. Everywhere evolution conquered, creation retreated. Men came from apes. In Britain Piltdown, in the orient Java and Peking man, proved it. In America?

In 1922 William Jennings Bryan, a successful politician who had in 1896, at the age of thirty-six, been nominated Democratic candidate for the Presidency, was campaigning in the courts against children being taught in schools that they were descended from apes.

On receiving a tooth, found in Pliocene deposits, from a Mr Harold Cook, H.F. Osborn (Head of the American Museum of Natural History) declared it had characteristics which were a mixture of human, chimpanzee and *Pithecanthropus*. He named it *Hesperopithecus haroldcookii* (Harold Cook's Evening Ape). In Britain Sir Grafton Elliott Smith fully supported him and, on the basis of this single tooth, there appeared, in the *Illustrated London News* (24 June 1922) a centre-spread of an artist's impression of 'man-ape' *Hesperopithecus* cavorting with his wife. Later investigation proved that the tooth was that of an extinct pig. So a pig made a monkey out of an evolutionist, though no publicity was given to the fact; although discredited, the misinformation had its insidious effect later.

The idea of the Scopes Monkey Trial of 1925 in Dayton, Tennessee, seems to have been hatched in New York by officers of the American Civil Liberties Union (ACLU). The legal defence, which hired famous criminal lawyer Clarence Seward Darrow, was arranged and paid for by the ACLU and members of the American Association for the Advancement of Science. The ACLU released to the Tennessee newspapers a call for a teacher who would break the 1925 state law against teaching evolution. The plans for John Scopes, a football coach and substitute science teacher, to be the defendant, were made at an informal meeting in a Dayton drugstore.

Whether or not he actually violated the anti-evolution law, Scopes was indicted and W.J. Bryan came down to act as special prosecutor. Darrow, from Chicago, acted as chief counsel for the defence. Two main lines of evidence for evolution were the Piltdown man and Nebraska man (*Hesperopithecus*). Nowhere in the trial did the scientific problems receive any sensible discussion. Darrow displayed ignorance both about the theory of evolution and the teachings of the Bible, and levelled a barrage of insult and vilification at fundamentalist Bryan. Bryan did not respond in kind. Darrow was clearly the media favourite, however. Scopes was convicted of violating the Butler Act and fined $100. His conviction was overturned by the State Supreme Court. Bryan, while resting in Dayton after the case, died. Creationism was dead. And, as the ACLU wished, the public was educated on evolution.

It is the same story today. The ACLU and fundamentalists are still locked in mortal combat: and bones still rattle and roll. 'Ass is taken for a man' (*Daily Telegraph*, 14 May 1984, p. 16). A skull, found in Spain and promoted as the oldest example of *Homo* in Eurasia, was later identified as that of a young donkey. In this case it seems man *was* an ass! In an article called 'Humanoid Collarbone Exposed as Dolphin's Rib'[8], Tim White, an associate of Don Johanson, accuses a fellow anthropologist of a *faux pas* on the scale of *Hesperopithecus* and Piltdown man. He does so because the bone in question is not properly curved and has a tiny opening, called the nutrient foramen, opening the wrong way for an ape-man. White has written: 'The problem with a lot of anthropologists is that they want so much to find a hominid that any scrap of bone becomes a hominid bone.'

On this note, let's get back to biology and see if truth gets a better deal among the cells, chromosomes and genes.

Notes

1   Transactions of the Victoria Institute, vol. 67, p. 21.
2   Von Koenigswald, G.H.R., *Meeting Prehistoric Man*, Thames and Hudson, 1956, p. 26.
3   ibid., p. 38.
4   Harrison Matthews, L., 'Piltdown Man' (a detective story in ten parts), *New Scientist*, 30 April–2 July 1981.
    Bowden, M., *Ape-Man, Fact or Fallacy?* Sovereign Publications, Bromley, 1981.
    Gould S.J., *The Panda's Thumb*, ch. 10, Norton, 1982.
5   Millar, R., *The Piltdown Men*, Gollancz, 1972, p. 232.
6   Boule, M., and Vallois, F., *Fossil Men*, Dryden Press, 1957, p. 141.
7   ibid., p. 145.
8   Anderson, I., 'Humanoid Collarbone Exposed as Dolphin's Rib', *New Scientist*, vol. 98, no. 1355, 28 April 1983, p. 199.

7 Seductive Trends

Without reproduction no organism of any kind would survive beyond a single generation. From the creationist point of view, reproduction is therefore a logical prime necessity. It is also central to the evolutionist point of view. In this case, however, the question arises: *why* is survival important? Why should genes – which are simply packets of chemicals – seek to preserve and complicate themselves, evolving all sorts of elaborate biological forms as vehicles for their protection and propagation?

There are three basic kinds of reproduction. Prokaryotic cells divide, but not simply! The single chromosome replicates, is drawn through a 'fork' and split in two. Daughter cells, each with a chromosome, then form by division. Eukaryotic cells can divide either by mitosis or by meiosis (fig. 8); that is, they undergo asexual or, in the special case of sex cells, sexual divisions. All the different reproductive strategies in the world are variations based on one of these three 'subroutines'.

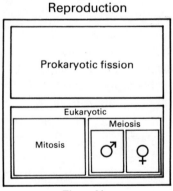

Figure 11.

How One Makes Two

No organism is fitter for its role in life than a prokaryote. In favourable conditions bacterial cells can divide every twenty minutes or so, giving rise, in theory at least, to 10^{20} cells in a day. In spite of these numbers, and the close attention of geneticists, no bacteria have been observed, in nearly a century of study, to turn into anything else. Most detailed studies have been made with the bacterium *Escherichia coli*, whose single chromosome or circlet of DNA is believed to carry about 5000 genes – the equivalent of a million three-letter codons. Even at this 'simplest' level of nature, replication of the chromosome in a bacterium that is about to split is a complex, exact business whose end-point is two identical rings where there was one before.

These bacterial rings contain DNA without any non-coding sections. Nothing is wasted. Indeed, if we examine the even humbler virus, $\phi \times 174$, which inhabits our bowel along with E-coli bacteria, we find startling economy. Scientists who unravelled its DNA found that there was not enough information to produce the proteins in its membrane. It now appears that sufficient information is found by the use of a frame-shift. In short, a gene is read off from the first base to produce protein A. Then the same message is read missing out the first base and starting with the second. This translates as protein B which fits with protein A forming the membrane of the virus. Try writing a message within a message. It is very hard. Such 'doubling up' indicates design of a very high degree. Higher organisms, including humans, do have 'wasteful' non-coding sections of DNA. Therefore, in this basic respect, we find the bacterium and a virus in advance of humans. Surely they did not, as Nobel-prizewinning molecular biologist Wally Gilbert once jokingly suggested, evolve from us!

In bacteria mutations have been observed which confer advantage in specialized environments e.g., the presence of antibiotics: but no new species, only new strains, have been produced. And neither the evolution of prokaryote to eukaryote nor mitosis from bacterial fission have been observed in any cultures. It is, though, sometimes claimed that primitive sex occurs in bacteria. In a few types small DNA loops called F (for fertility) factors can sometimes become established apart from the main chromosome. In conjugation a 'have' can form a slender tube which penetrates the cell wall and injects a 'have not' with part or all of an F factor. 'He' then dies and 'she' becomes a 'he'. After transfer the extraneous DNA, which is of dubious importance and generates no non-bacterial feature, is usually degraded or expelled.

Bacterial conjugation resembles the way a virus infects a host, which no one calls sexual: on the other hand, it is a haphazard, imprecise process which bears no resemblance to the purposeful patterns of mitosis or meiosis. Prokaryotes have a precise and efficient method of reproduction.

Only the theory of evolution demands that we cast the transfer of F factors, like a bizarre genetic disease, in the role of a sexual act.

Eukaryotic cells have more than one chromosome, with genes built up from segments in ways that make them quite different from the simple DNA of the prokaryote. Could the one kind of chromosome have developed from the other? Not a chance, say modern biologists: they differ radically in structure and function, and this difference may allow eukaryotes to change and diversify — as prokaryotes cannot and never will.

Consider *mitosis* more closely (fig. 8). In eukaryotic organisms where at least two chromosomes (and therefore four chromatids) are present, some apparatus is required to ensure the units are separated so that both chromosomes are represented on each side. With higher numbers, say fifty chromosomes, it is essential that muddle be systematically avoided and each new cell contain its full complement of twenty-five homologous pairs. This necessity is achieved with masterly economy. The chromosomes shorten and thicken as the result of extensive coiling. In this way they can move without getting tangled up. They lie along the equator of the spindle's web, attached by a specialized single region, the centromere, which 'knots' each two sister chromatids together. Next, the centromere doubles and the chromatids are drawn apart, each towards a pole of the spindle. When this stage is complete, the cell membrane constricts and eventually pinches off, forming two cells: in the case of plants, a cellulose cell wall forms between the two halves. The job is done, the apparatus dismantled. The product of this dance in time and space underwrites eukaryotic growth, cell replacement, regeneration and asexual reproduction.

Multiple fission, as in the malarial parasite or in protozoan spore formation, occurs when more than two daughter cells are generated. More startling than multiple fission is fragmentation, which occurs in animals with great regenerative powers, such as sponges, hydroids and flatworms. Fragments of non-parasitic flatworms will grow into new individuals; the single worm *Linaeus* can regenerate hundreds of new individuals from fragments of itself.

The asexual theme, or subroutine, is varied in budding where an outgrowth from the parent usually detaches and grows into a self-supporting adult. Distribution of this subroutine is irregular, appearing in hydra, some flatworms, sponges, segmented worms, seasquirts etc. It occurs in fungi (yeasts). In plants it is called gemmation, and is common in mosses and liverworts. How, with such a distribution, is it supposed to have evolved? Or did it evolve many different times?

When part of a plant body becomes detached and develops into a new self-supporting plant, vegetative reproduction is said to occur. Rhizomes, bulbs, tubers, corms and runners are variations upon this theme and witness to the importance of a two-pronged reproductive drive in organisms that cannot move to meet a mate. The other prong is sporulation. In plants and fungi, as well as in protozoa, a wide range of spore types are dispersed to pioneer new lands or to endure hard conditions.

The mechanism and associated biology of the process may vary slightly from organism to organism, but the overall concept of mitosis is universal in eukaryotes, underwriting the rejuvenation of cellular life. Its mechanical means of information transfer is, argues the creationist, a circumstantial necessity where any more than one chromosome is gathered in the nuclear heart.

How Two Make One

Where parental reproductive cells fuse, forming a fertilized egg cell, two make one. In biological terms, the syngamy of two gametes produces a zygote. Here the reciprocity, polarity and conjugation of the sex theme are fully expressed. Male and female, dual aspects of a single sort of organism, unite to recreate themselves.

Sexual reproduction represents a flowering, a sharing, a recombination. At its centre is meiosis, the complex double division of gene information (fig. 8). On the basis of economical and efficient design, it might be expected to be the same process in man and monkey, moss and maize, and so it is. The goal of meiosis is to halve the duplicated genetic information available in the cell to form two haploid gametes which, reunited, engender a new diploid individual.

When, at the start of meiosis, chromosomes appear in the nucleus, they are arranged in homologous pairs (bivalents), side by side, one from each of the cell's parental contribution. A molecular embrace called crossing over (fig. 9) occurs between chromatids. This rite is also called genetic recombination because it involves the breakage and exchange of material between chromatids. New permutations of characters can be expected but no *new* material is generated.

The dance continues in an orderly fashion. The chromatids part, split longitudinally, and the halves are drawn into different ends of the dividing nucleus. The result is a gamete, a reproductive half-cell ready to fuse with a counterpart from another organism of the same species, but unique in its assembly of characters.

For the creationist, meiosis bears the unmistakable hallmark of planning. It is a pre-programmed routine designed to extract maximum variability at minimum costs in labour and materials. It underwrites the sexual reproduction of some single-celled and nearly all many-celled organisms.

The Origins of Sexual Reproduction

If you believed that 'selfish' genes were out to propagate themselves in the most prodigal way, you might be surprised that meiosis ever occurred. Why? Because only half the genes of an individual would be transmitted and, therefore, they would spread only half as fast as in asexual rivals.

But sex is successful. You might therefore argue that meiosis is the chief instrument of evolution which introduces variety and ensures that one individual always differs from another: in this way, it underwrites evolution. This is a common but fallacious argument. Sex is for *shuffling* (fig. 9). Certainly, quite different hands of cards can be dealt from the same pack. A hand can be mixed, or it might be all hearts or all clubs. In certain circumstances, certain permutations may excel: the parable of the coloured birds (p. 77) showed this. But it does not mean that the organism has, in any sense, evolved. Sexual recombination creates variation but does not amount to evolution, whose sole neo-Darwinian differentiator is mutation. So why do Miss Greece (Venus) and Mr Universe (Apollo) look so human yet divinely different? What is the origin of sex?

We do not know. We simply do not know how bacterial fission evolved, though such a life-support system must have been active and accurate from the start. Nor have we much idea how mitosis arose: nor meiosis (presumably) from it. Specialized equipment, such as the spindle, linear chromosomes and centromeres, are required for their electro-mechanical choreography. How did the dance take place without them? Or before they were functional? Prokaryotic division and mitosis produce clones – cells from a single ancestor which have the same genetic makeup. The meiotic shuffle, a dance with extra steps, is creative as well as procreative: it helps to vary the successor cells (called zygotes) from which unique clones (individuals like you and me) can develop. As well as extra steps, meiosis needs extra equipment. It requires not only the curious fission-dance of chromosome and chromatid, and the even-more-curious wrapping and part-exchange of genetic material, but the whole elaborate mechanism of sex – maleness and femaleness in their myriad manifestations throughout the plant and animal kingdoms. How did all this come about? It is worth having a look at some of the simpler plants and animals in order to trace the origins of sexuality which, like the origins of mitosis and meiosis, are far from clear.

Sex, everyone is agreed, generates the spice of life − variety. Sexual apparatus is a delivery and back-up system for the central mechanisms, meiosis and its reverse, fertilization. Of course, the programme for male and female forms must already be written and, in each respect, correct before sex can function. No use one evolving before the other!

According to the theory of evolution, life on earth endured no 'adolescence'. Sexual maturity, in algae, jellyfish and other forms, was present in pre-Cambrian seas. Land-plants and animals incorporate a rich variety of sexual parts and techniques. Some green algae and fungi have a sexual phase in which they produce 'isogametes', indistinguishable as 'male' and 'female' except that they will only fuse with other isogametes from a different plant. So they are called, sexlessly, plus and minus gametes. In most cases, however, there is a clear difference between male and female gametes − the smaller male tends to be active and the female passive.

However 'simple' sex is, one thing is clear. *Achyla* is a fungus whose body is composed of filaments which spread by growth at the tip and by branching, into a network called a mycelium. When lying side by side in water, couples of some strains of *Achyla* develop, in one, male sexual organs and, in the other, female. The fully developed male organ will penetrate the wall of its partner's female organ and release sperm cells which fertilize the eggs that have been formed inside. Several different chemical substances are involved in the mating procedure. As we extend our study to seaweeds, mosses or even human beings, we learn that the sex system is based on an exact programme of chemical switches and signals − mostly in the form of hormones. Not only the correct genetic subroutines and sex organs but the right chemicals in the right sequence are required for satisfactory sex.

In protozoa, sexual reproduction is common only among parasitic forms, and in some of the most complex of all single-celled organisms, the ciliates (for example, *Paramecium*). Ciliates usually just divide in two, in the normal unicellular way. Eventually, however, strains weaken and die unless plus and minus individuals conjugate after a series of divisions involving one of their two nuclei. The process is complex enough to deter any evolutionist from the claim that sexual reproduction arose among *Paramecium* and its kin.

In life-cycles where sexual and asexual forms alternate, the two forms are often very different. Take a jellyfish. The form that stings is the sexual medusa; in its background is an alternate generation − a tiny asexual polyp that lives close to the sea bed and buds off medusae one after another. Both jellyfish generations are diploid, having a full complement of chromosomes; but in, say, ferns, fungi and algae one phase is haploid, with half a full complement, and the other phase is diploid. Whatever the case, it is clear that both forms must contain genetic information for the whole cycle, although only a part of it is used to generate the sexual body. Did sex evolve in each case separately, as a second string to an asexual bow? If not, where did it come from?

For the evolutionist bodies are merely gene-capsules, varied vehicles for the determined, selfish survival and replication of a remarkable chemical, DNA. The advantage of sex, whose origins are unknown, is the production of evolutionary novelties.

Contrast this with the creationist view. Here genes are secondary: they are as important as the score in a musical work – vital for transmission of the art but secondary to the realization, in real music, of its potential. Genes, then, are chemical symbols. They represent the spoken language of life. And sex is the conservative agent whose object is to maintain but vary the archetype: to populate a changing world with as many diverse and healthy individuals as possible.

Sir Clive Sinclair, a leading British inventor, pioneered the digital watch, inexpensive calculators, computers for the ordinary citizen and flat-screened mini-televisions. Turning his attention to the development of an electric car, this creator favoured a 'modular' approach to vehicle design – the idea that a basic chassis can be designed to take different types of body style – a single-seater car, a family shopping car or a delivery van. For the creationist, DNA subroutines underlie biological modules. This applies, as to other homologies, to reproduction and to sex. The variety of reproductive equipment and techniques are, the creationist believes, best explained in this way.

'Half sex'

Parthenogenesis, a curious sideline, is the development of an ovum (in plants, an ovule) into a new individual without fertilization – in a sense, 'half-sex'.

Such diverse types as dandelions, wingless aphids and maleless water fleas are often the product of parthenogenesis. So are some turkeys, wasps and all drone bees. Artificial parthenogenesis, by pricking the egg, has produced toads. Virgin birth, however, such as the union between a god and a woman involving n human and n divine chromosomes, has not been scientifically recorded in higher species!

When parthogeny is possible, why should sexual reproduction continue to exist? After all (and feminists may concur) males represent a two-fold burden on females. If, for every male required to consummate further sexual increase, a female could produce another female parthenogenetically, the population would double quickly. Such females would not have to wait for a male in order to reproduce; why, for fitness' sake, waste energy and resources on producing males?

Professor John Maynard Smith, who puts forward this argument, concludes that short-term individual gain in parthenogenesis is outweighed by long-term impoverishment of the species, because it will not evolve to overcome changes of environment.[1] Bacteria lack recombination, however, and do very well. Perhaps sex has short-term benefits such as pleasure and (at least in higher species, which does not explain its origin) parental care. Possibly it has an ecological role, as with *Spirogyra*, water fleas and onions, generating various forms of zygote in harsh conditions.

Where is the pattern? No one suggests that virgin birth is a halfway stage or even a real competitor to sexual reproduction; nor that the female of the species preceded the male. In all parthenogenetic species, which are restricted, there is a parallel sexual type. Perhaps it marks a degeneracy, as in Lady's Mantle where the stamens have become vestigial, some being lost entirely. If present, they are either empty of pollen or contain grains whose sexual potency is lost. This suggests that, although the coding for a complete organism resides in a haploid set of chromosomes, parthenogenesis is, except in rare instances such as drone bees, an aberrant condition and offers no solution to the problem of the origin of sex.

'Double sex'

Hermaphroditism is another curiosity in the story of sexual reproduction. Hermaphroditus was the son of Hermes and Aphrodite, so called because he and the water nymph Salmacis became one person. Rather than parthenogenesis, the creationist prefers that hermophrodites may, in their bisexual form, reflect the true inward basis of sex.

Many plants and animals are hermaphroditic, i.e. each individual has both male and female organs. When flower petals attract pollinators, mechanisms exist which stop 'selfing', i.e. inbreeding. Usually male and female parts of a single flower will mature at different times. Even if self-fertilization does occur, outcrossing is also possible; and it is commoner for wind- or water-pollinated flowers to be single-sexed.

In contrast, hermaphrodite animals are scattered without order across the invertebrate phyla. Some snails, sea lilies, bivalves, segmented and ribbon worms, corals, endoprocts, lamp shells, sea anemones, Crustacea and bony fishes such as porgies, picarels and rainbow wrasse are hermaphrodite. So are complete taxa of flukes and tapeworms. It seems that a hermaphrodite might have more of a chance in conditions where mates are scarce. It can 'self'; alternatively, all the creatures it meets, and not just 50 per cent of them, will be potential mates. There is not much information available about self-fertility in hermaphrodites. Certainly in earthworms elaborate precautions are taken against self-fertilization; in other creatures, successive hermaphroditism means that, as in plants, male and female organs are active at different times.

If an organism contains 'weak' or mutant forms of genes, these are usually overridden by healthy alleles. An inbreeding type runs the risk that its unhealthy alleles will be duplicated, not overwritten by dominant genes from outside; of course, unhealthy homozygotes will realize an unhealthy physical form. It seems, therefore, that the mechanisms and instincts to outbreed are programmed as part of the package into all sexual organisms.

An intersex form is of certain interest to the creationist. Such an individual, whose sex-determining chromosomes or sex-hormonal controls have failed, is intermediate in character between the two sexes and may, or may not, be hermaphrodite.

Sex as a whole

The creationist's interest in the development of sex is based on a belief that all forms of life that exhibit sex are conceived as one whole, and divided into unisexual halves. The programme, like the unified Yin-Yang symbol (p. 110), expresses the duality of complements within a unity. For the psychological development of this idea, the reader can consult Taoist literature such as the *Tao Te Ching* or *I Ching*. Biologically, the DNA coding is hermaphrodite, enabling any sexual organism to become male or female except that, during development, the organism is triggered into one channel or the other. The code is hermaphrodite, but its expression is in a single-sex body.

This concept of sex does not require the simultaneous evolution of meiosis, male and female sexual organs and instincts. It requires no 'hopeful monsters', searching at every stage of evolution for another 'hopeful monster' of the opposite sex. Instead, sex in plants and animals is the product of precise, purposive programming. From algal conjugation to the conjugal arts of man, it represents a central reproductive theme around which bodies are tightly woven. Above all, it requires that *coding for both sexes* be present in every individual of a kind, right from the start. We can consider both plants and animals in the light of this viewpoint.

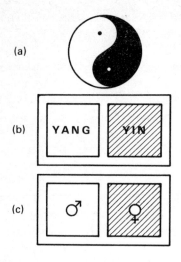

Figure 12. **The Unity of Opposites.** According to the philosophy of ancient China, the cosmos is an expression of Tao. The Tao is one: it is infinite, unformed and timeless. The evident, finite, time-bound articles of the cosmos are derived, according to a principle of 'opposites', from the mysterious Tao. A dynamic interplay of 'opposites', such as light and dark or male and female, is supposed to pervade every facet of the universe. Every object expresses, in different proportions, various yin or yang characteristics. In Christian terms, Tao is the Word, by and in which the cosmos was created and is sustained. The T'ai-chi T'u symbol neatly encapsulates cosmic duality within the unity of its Source, the Tao.

(a) *Ancient Chinese yin-yang symbol*, called T'ai-chi T'u or 'Diagram of the Supreme Ultimate'.

(b) *A Chinese box version of the yin-yang symbol.*

(c) Yang (male) and yin (female)

Sex and the single plant

Evidence for the origin of botanical life is discussed more fully in chapter 12, but now is the place to mention both the Hofmeister series and some of the tricks plants play to get fertilized – although why they bother with sex when that fail-safe device, vegetative propagation, works so sturdily, is another question. Bulbs, tubers, runners and rhizomes are surely simpler than the refinements sex employs.

In their life-cycle plants show an alternation of generations. This is seen most clearly when the two generations are independent. In ferns and mosses, for example, the green plants we know are the sporophyte, asexual generation. They are diploid and produce spores which develop into a tiny gametophyte, almost too small to see without a lens. This bears male and female gametes – haploid sperm and ova – and one of each fuse to produce the diploid embryo that grows into another sporophyte. A 'primitive' life-cycle is often considered to be one in which – as in mosses – the gametophyte is as large as, or larger than, the sporophyte. In 'advanced' plants the sporophyte is much larger than the gametophyte and may even contain it.

How does this work in flowering plants? The rose bush, grass or tree is considered the equivalent of the sporophyte generation in mosses and ferns. The tiny male and female parts of the stamen and carpels that produce, respectively, pollen grains and ova, represent a much reduced gametophyte generation.

From mosses and ferns to flowering plants we can find a series (botanists call it the Hofmeister series) with gradual reduction of the gametophyte – which is large in mosses, small but separate in ferns and internalized (within the sporophyte) of flowering plants. Conversely, the sporophyte in flowering plants is large, containing the gametophyte in its flower; large but separate, as fronds, in ferns; and small in mosses. In the animal kingdom we find an analogous series wherein offspring of the so-called 'primitive' animals develop independently, while those of 'advanced' animals become 'eventually' internalized within the mother. It stretches from hydra, with its alternation of generations, through invertebrates and insects, fish, reptiles, birds and then monotreme, marsupial and placental mammals.

It is tempting to perceive an evolutionary trend both in the internalization of embryos and the Hofmeister series. Both series are striking, but there is no real evidence that either has anything to do with evolution. Algae show all sorts of life-cycle, 'primitive' and 'advanced'. *Spirogyra* reproduce sexually and have no real sporophyte generation: but another green alga, *Cladophora*, has only one generation – the sporophyte. Another kind of *Cladophora* has complete alternation of an independent sporophyte and gametophyte generation. In the brown seaweed, *Fucus*, the gametophyte generation is as insignificant (yet important) as in flowering plants; but its operation, with antheridia and female oogonia, is scarcely less complex, less suitable. Unlike *Fucus*, sporophyte and gametophyte generations of the sea lettuce (*Ulva lactuca*) look identical. They differ only in that motile spore cells have four flagella whereas motile gamete cells have only two. Such a spread of life-cycle patterns within the algae were better termed 'different' than 'primitive' or 'advanced'. Mosaic rather than evolutionary, they appear as evidence in favour of created permutations.

Of course, any trend towards complexity would involve the evolution of stem, root, leaves, the vascular system with its water relationships, and dispersal by seed instead of spore. It would rise to a peak in the sexual beauty of the angiosperm flower. In fact, land-plants appear with the basic architecture and major features of their life-cycle – except flowers – complete in half a single geological age (mid- to late-Devonian). The seductive Hofmeister trend begins underwater, passes up through plants that require moist conditions for their reproduction and culminates in flowering plants, including grasses and trees. But algae, like the others on land, have their crucial role to play in the network of life – in water. It seems reasonable for the creationist to insist that the 'trend' is no more than varied subroutines, permutations of the reproductive modules necessary to ensure the transfer of sex cells in wet and dry conditions.

Self-fertilization leads, like incest in most cases, to reduced vigour. Since 1676 when Sir Thomas Millington, F R S and the botanist, Nehemiah Grew, introduced a puritanical world to the idea that flowers are sexual, study has revealed that these bisexual organisms employ all sorts of tricks to ensure that their pollen reaches another plant.

The wind can effect inaccurate dispersal: there is, however, no evidence that the naked and exposed sexual organs of wind-pollinated flowers preceded those pollinated by insects. Indeed, flowers may have thrown their sex lives to the wind later rather than sooner. The stinging nettle is wind-pollinated but it has, in its flower, glands designed to produce nectar. Some plantains are pollinated by wind, others by insects. The field poppy is, for its part, insect-pollinated but produces more pollen than most wind-pollinated plants. The argument that such pollination is 'primitive' or 'wasteful' is fallacious. Many highly developed forms of life produce enormous amounts of sperm − in humans hundreds of millions of spermatozoa compete to fertilize a single egg. And, in a broader ecological context, many other organisms benefit from pollen (and animal eggs) as a source of nourishment. Wind pollination is supposed to have appeared late in the history of flowering plants.

Accurate pollination is achieved through the employment mainly of insect pollinators. When the young and attractive female part of a flower is ready to receive the male, signs of sexual arousal appear. Secretion from 'her' tissues prepares the way of intercourse in a way that resembles that of an animal counterpart when ready for mating. Heightened colour, attractive scent and moistening of the stigmatic surface with a sugary substance occur. Chemical analysis shows that floral perfumes are like insect body odours and act as stimulatory signals. An insect is drawn by sight, smell and, from the nectar and pollen, the promise of food and sensations of taste. It may even be deceived by shapes, patterns and colours which resemble, as in the Bee Orchid, a female of the species with the same markings that guide the male bee towards her genital area. Pseudo-copulation may ensue, until the frustrated male leaves to try another flower, pollinating it in the process. The mimicry used by flowers to trick insects into their service is remarkable, versatile and, in many cases, defies any adequate evolutionary explanation.

Different types of stigma and pollen ensure that the female receives only the male of her species. It even happens that, as pollen grains on the stigma compete in driving pollen tubes towards her ovules, she may secrete hormones which diffuse into her tissues to discourage incompatible males and encourage 'foreign' pollen from another plant of her type. If you inquire why her own flower's pollen is normally excluded, you will learn that either male and female parts are physically isolated from each other's influence or they mature at different times, so avoiding the unwanted 'incest'.

There exist further resemblances between flowers, at the 'top' of the plant kingdom, and vertebrate animals. Apart from the similar production of sex cells in anthers (like testes) and ovaries, flowers also have a placenta, the part of the ovary wall on which ovules are borne. An ovule is attached to it by an 'umbilical cord' called the funiculus. A scar called a hilum is left when, as the seed ripens, the funiculus breaks: this is the equivalent of our navel. In plants transfer cells direct solutes rich in sucrose and mineral ions to places, such as the root and shoot of a seedling, where they are needed. In a seed there exists a striking parallel between such cells, as they transfer nutrients from food-storage tissue called endosperm to the plant embryo, and cells of the mammalian placenta. Both seed and womb nourish an embryo. The creationist finds it extraordinary that such analogous reproductive equipment should have evolved by accident in the two separate kingdoms. For him, bisexual flowers simply express the *whole* of the same plan which is expressed as two sexed halves in vertebrates.

Furthermore, man lies locked in an energy-cycle with plants (chapter 10). Just as reproductive similarities can be observed, so a close and complementary relationship can be drawn between photosynthetic and 'reverse photosynthetic' (respiratory) systems. The Yin-Yang concept of interlocking opposites applies. Yin is 'female' and Yang is 'male'. The male, whether by pollen or by sperm, acts on the receptive female: she produces. Life depends upon her fruit; in her womb, not the man's, lie the seeds of future generations. Perhaps, as he observes it inveigle pollen and bear fruit, man rightly intermingles the two kingdoms and sees in a flower the metaphor for woman. And, he may wonder, is it an accident or was it rightly planned?

Animal sex

If there exist similarities between the sexual structures of a plant and animal, there exist still closer links between male and female. As the creationist sees it, the 'hermaphrodite' plan for *human being* would remain constant although, in practice, differentiated into a male or a female individual. In the interests of economy, male and female characters would develop as modifications of common embryonic tissues and structures.

Human sex is codified by a small and economical genetic difference in just two out of the forty-six chromosomes in the zygote. Two chromosomes carrying an X factor will result in a woman; an X chromosome and a Y chromosome will result in a man. This microscopic switch shunts development into alternate pathways that use the same basic materials to produce different end-products. The knot of tissue that under XX guidance becomes an ovary, under XY becomes a testis: the folds of skin that in the female become labia, in the male fuse to form a scrotum. Clitoris and main shaft of penis, uterus and prostate are similar alternatives, triggered from the same elements but developed differently by the chromosome switch.

Although possession of two X factors or an XY combination seems to act as the primary sex-switch, genes essential for male development are located on the X as well as the Y chromosome. The pituitary gland of both sexes secretes the same hormones to control the activity of the gonads. They are called luteinizing hormone (LH) and follicle-stimulating hormone (FSH). The gonads themselves release steroid hormones such as oestrogens, androgens and progestins. These are manufactured by a common route, from cholesterol through progesterone to either androgen or oestrogen. These 'male' and 'female' molecules are interconvertible. Therefore, although the action of testosterone promotes 'maleness', traces of it can be present in the ovary. Conversely, although oestrogens promote 'femaleness', quantities may be found in the testes of boar, stallion or man.

It is a question of bias and balance. In the human embryo, no recognizable sex is apparent for six weeks after fertilization. Thereafter it is determined by the differential secretion of steroid hormones from the gonads. If the gonad secretes testosterone, a male reproductive system will result. In its absence, a baby girl develops. In fact, when female embryos are artificially treated with testosterone, they engender both male *and* female gonads and ducts!

Later, at puberty, the action of the same gonadotropins (LH and FSH) on either boy or girl generates a man or a woman. And the presence of testosterone at birth is thought to 'sex' the hypothalamus in the brain; this organ, now male, later, suppresses the mechanism that causes female 'periods'.

Take nipples. Every male mammal possesses rudimentary nipples, as well as other feminine organs in rudimentary form. How shall we interpret them according to neo-Darwinian theory? If both male and female forms were originally rudimentary, what selective advantage was there for a milkless milk-production system to gradually improve to the point, a few million years later, of working? Conversely, if the male nipple is 'rudimentary', does this imply that, in the past males suckled?

Other important steroids besides those for sex and breastfeeding occur in bacteria, algae, higher plants and animals. All steroids are transformed from cholesterol down a line of specific enzymes. On this line 'male' precede 'female' hormones. Do we conclude that 'male' anticipated 'female'? There is no evidence whatsoever for the prebiotic origin or evolution of steroids: for the creationist they are simply molecular archetypes that work important chemistry in every grade of life.

It's funny! A clown-fish starts life as a male and then, if it becomes the largest fish in its group, becomes female. She is mated by the next largest fish, a male with larger testes than the rest. If this male is removed, the next largest develops big testes and becomes sexually active. What causes these dramatic sex changes is not known but, clearly, the genetic basis for either sex is present in every clown-fish.

Whether we study the phenotype, the organs or the biochemistry of sex, it seems that a *sex concept*, with its variations upon theme, better explains the facts than haphazard evolution, whose every step is unexplained. Bringing a cybernetic approach to bear on the study of sex, the creationist concludes that this intricate, sensitive network of feedback and control is not the product of a series of mutations! An apparently trivial control – the chromosomal switch – can initiate the different developmental paths that lead to male and female individuals. This surely implies that the end-product was accurately conceived. Further, it implies that the sex-concept (call it idea or archetype, if you like) was 'packed' into code in a way that, once triggered, would allow its own unfolding and accurately express a meaning. Whatever else philosophers, poets or lovers may discern, this 'meaning' is at least the growth and maturation of a specific life-form.

Sexual permutations

Darwinians seek to trace the origins of sex in another way. For them bacterial conjugation represents the simplest form of sexual reproduction; plus and minus strains in algae and fungi a slight advance; sex with external fertilization in water the next stage; and internal fertilization on land the ultimate advance. The trends from simplicity to complexity, from water-based to land-based forms and, generally, from small to large, are consummated in plants that flower and in animals that are born live and suckled – the mammals.

The creationist's interpretation of this sequence is that reproductive structures incorporate, in their DNA coding, permutations of sub-routines. The supposed Darwinian trend represents no more than the pre-programmed capability of different organisms, large and small, living in water or on land, to survive the physical and ecological constraints placed on them. We can examine these contrasting points of view in relation to reproduction among vertebrates and see which provides the clearer overall picture.

Oviparous animals lay eggs in which the embryos are undeveloped. Most invertebrates, fishes, many reptiles and all birds are good examples. Ovoviviparous animals have embryos which, although they develop within the mother, are separated from her by persistent egg membranes: many reptiles are ovoviviparous, so are one or two snails, some roaches, flies and beetles, and parthenogenetic aphids, gall-wasps, thrips etc. Viviparous animals have embryos which develop within the mother, in close contact with her through a special organ called a placenta. Nearly all mammals are viviparous.

These three conditions are held by orthodox Darwinians to represent evolutionary stages. For oviparity, no developed male organ is required; nor any genital contact in fertilization. In the second and third cases juxtaposition of male and female organs or penetration by the male is necessary; fewer eggs are required because there is greater certainty of fertilization, and less chance that eggs will be lost. Also, it is supposed that greater physical contact in mating heralds increased parental care for the fewer offspring that are born.

As everyone knows, mammals are higher animals than fishes, so mammalian viviparity must represent an advance on fishy oviparity. But there are many exceptions. Sharks and their kin, the dogfishes and rays, are supposed to be primitive fishes, and should be oviparous. True, dogfish lay eggs in horny cases: they become attached to seaweed and we call them mermaids' purses when we find them on the beach. Most species of sharks and rays, however, are viviparous. The female, having been internally fertilized by two male organs called 'claspers' (a kind of double penis) gives birth to up to sixty live pups. A newborn shark resembles the adult in almost every way. Sand-sharks have an interesting variation on viviparity – of a hundred or so eggs stored in the female's oviduct, only the first two hatch. They support themselves by eating all the other eggs, emerging about a year later, viviparous and almost half their mother's size. Many gradations between ovoviviparity and viviparity can be traced in this group. Some rays are oviparous, others are viviparous, bearing up to twenty alevins at a time. Some even generate uterine milk, secreted by the oviduct, in addition to nutritious egg-yolk.

Some guppies too are viviparous and the discus fish, a cichlid native of the Amazonian basin, appears to feed its young with a sort of milk. Just before breeding a slimy protective coating on the scales of the adults thickens considerably. This is due to a copious whitish secretion, granular in composition, which changes into tiny filaments when pulled or rubbed. After absorbing their egg-sacs the youngsters swim directly to the parents where they cluster for protection and nourishment.

The Mediterranean cardinal fish lays her eggs in the mouth of the male who rolls them around, aerating them, until they hatch. This is reminiscent of 'marsupial' sea-horses: during courtship the female actively pursues the male and deposits her eggs from an orange 'penis' into a pouch on his belly where they are fertilized, sealed and nourished for six weeks on his blood. The 'pregnant' male then enters labour and two or three hundred young sea-horses are 'born'. Oviparity? Viviparity? It is hard to say: sea-horses reproduce in a way that cuts across our ideas of gender.

Oviparity, ovoviviparity and true viviparity are found in frogs and toads. Male Darwin's frogs rear their eggs and tadpoles in their mouths. Male Surinam toads help to attach the fertilized eggs to the backs of the females, whose rough skin swells and encloses up to sixty eggs. The young toads emerge from beneath the mother's skin.

Caecilians, which superficially resemble large earthworms but are amphibians, have species that lay eggs and others that are viviparous and produce uterine milk. In many salamanders the young may develop ovoviviparously. The olm, a Yugoslavian salamander, is oviparous and viviparous, and the black salamander is viviparous. The young are nourished, as in sharks, on unfertilized eggs in the oviduct.

Among reptiles skinks, lacertas, boas and vipers belong to groups that have both oviparous and viviparous members. Some species, the adder and common lizard, for example, lay eggs in warm parts of their ranges but, northwards, where it is colder, bear their young live. In some cases, while the young break out of the membrane before they are expelled, there is very little difference between viviparity and ovoviviparity. The three-horned chameleon is viviparous. So are reptiles living in the ocean like sea-snakes and, of yore, the ichthyosaur. The smooth snake is ovoviviparous and the sand lizard is a true egg-laying species. At least two kinds of lizard are parthenogenetic. In other kinds females preponderate, perhaps due to the presence of parthenogenesis. Two kinds of hermaphroditic lizard occur. It is not easy to account for this sort of mosaic distribution in evolutionary terms. But it is a natural consequence of the idea that three equally good but different types of parturition are coded into reptiles, only one of which is triggered in any individual.

Birds without exception lay eggs. Mammals are all viviparous except for one small group, the monotremes, which lay eggs. These grow as they pass slowly down the oviduct, emerging about the size of a sparrow's egg.

Marsupials, which were long thought of as 'primitive' mammals, in fact have a most efficient kind of viviparity. In some ways it improves on that of the so-called higher, placental mammals. At any time a female marsupial can nurse a large infant or Joey in the pouch, also carry a tiny newborn infant in her pouch, and have in her uterus an even tinier reserve embryo, just a few cells big, that starts to grow as soon as the Joey leaves. After expulsion as a blind, pink scrap of flesh no larger than a bean, the neonate hauls itself by instinct several centimetres across its mother's fur to the opening of her pouch. The next oestrous cycle will very likely bring the production of another embryo to wait in the wings for its place in the pouch. The marsupial production-line is no less 'fit' than the placental, in which one or a few embryos are reared internally from a single cell to full term. It is just different.

Why would you not expect to find a marsupial whale or a bird weighed down with an embryo chick in its womb? Apart from obvious cases like this, plants and animals incorporate different permutations of the main reproductive methods. Of course, some are biochemically and physiologically more complicated than others – in fact all methods are very complicated – but this is no reason to interpret them using subjective values. For bacteria, fission and not oviparity is suitable: this does not mean the simpler method is 'ancestral'. On the seashore there live oviparous, ovoviviparous and viviparous kinds of periwinkle. Land-dwellers usually employ different schemes from those used underwater: this does not demonstrate that one scheme is more or less 'advanced' than the other, or evolved from the other – but that organisms are suitably equipped to reproduce, and thereby survive, according to their own particular size and sort of dwelling-place.

You can argue (and the creationist would rebut your bio-logic) that in the beginning there was a Genetic Mistake. After this, as the result of many such mistakes, there evolved meiosis, sperm eggs, vehicles and the goal of sex, fertilization. This, however, is only the start of the amorous affair. Is the development of an embryo the product of a series of mutations? Did birth, then adulthood come from genetic accidents?

Notes

1   Smith, J. Maynard, *The Evolution of Sex*, Cambridge University Press, 1978, pp. 2, 3 and chap. 4.

8 Wrong from the Start − or Right?

The transfer of genetic material in the cause of reproduction is one thing; the development of new forms is the next stage. It is to this that we now turn.

Growing out of the Past

Pick up an acorn and examine it in the palm of your hand. Shining beauty it may possess, but it also encapsulates technology far beyond the reach of any laboratory. From little acorns mighty oak trees grow! Within this acorn resides the potential for trunk, leaves, transport networks, flowers and thousands more acorns, each yet another potential tree. All this is packaged as precise, pre-programmed information, in a manner which makes advanced microchip technology seem positively gross. The programmes are keys, not only in space but also in the dance of developmental time − keys to the poem, the symphony and the science we call an oak tree. There are no mistakes: this acorn will not become a beech, a hawthorn or a horse. You developed just as surely from an egg smaller than a full stop on this page; the instructions it contained were two billion codons, equal to the verbal information contained in 2000 encyclopaedias, each 1000 pages long with 1000 words on each page. And was that all?

In 1828 Karl von Baer (1792−1876), the founder of embryology, published hypotheses based on observations of the similarity of embryos of mammals, birds, lizards and snakes during their earlier stages. In the early development of the egg, he said, general characters appear before specialized ones: and the specialized ones develop from the generalized ones. During the development of an animal there is a progressive departure from the form of other animals. In this way, he noticed, the early stages of 'higher' animals resemble the early stages − but not the adults − of 'lower' animals.

This makes sense. When a sculptor starts to carve a block of wood into a model, he immediately reduces its potential. The more he carves, the fewer possibilities remain until the unique model he was aiming for emerges. Much the same happens in embryology. A growth theme seems to operate. Triggered by the spermatozoa a chemical dance − a *strictly programmed* dance − begins; each figure in the dance restricts possibilities for future figures and the dance, like any other metabolic process, proceeds to a definite and determined end.

The analogy with dance and music is not idle. With perfect timing and adherence to score, the DNA orchestrates a symphony of protein inter-actions. A unique 'musical' shape, reaching full expression in the adult, develops. Or, to use the computer analogy, a complex three-dimensional stereo-computer programme is run through to its pre-set conclusion.

To its *pre-set* conclusion. Cells come from cells: parents generate only their own kind and here development differs firmly from evolution. Both development and evolution represent increases in complexity from simple starting-points. But development is precisely coded and targeted: evo-lution is open-ended, goal-less, with no recognizable end-points. The development of the individual from an egg (ontogeny) is a different kind of process from the evolution of an individual from hypothetical ancestors (phylogeny). The distinction is especially important to creationists, who have no difficulty in accepting the one but doubt if the other − phylogeny − is more than a figment of the evolutionist's mind.

Nevertheless, the two are often coupled and even confused. In 1864 Fritz Muller reinterpreted von Baer's work in an evolutionary way, and in 1868 Haeckel used this reinterpretation to formulate his 'fundamental bio-genetic law' or recapitulation theory. This stated that an organism, in the course of its ontogeny or embryonic development, successively passes through (recapitulates) phylogenetic stages passed through by its ancestors. That 'ontogeny recapitulates phylogeny' became a popular notion among biologists. Fired by enthusiasm, Haeckel stated that the entire animal kingdom was descended from an organism resembling the gastrula − an early stage in the embryonic development of most animals. To support his case he began to fake evidence. Charged with fraud by five professors and convicted by a university court at Jena, he agreed that a small percentage of his embryonic drawings were forgeries; he was merely filling in and reconstructing the missing links when the evidence was thin, and he claimed unblushingly that 'hundreds of the best observers and bio-logists lie under the same charge'.

Haeckel's main fault lay not, however, in his cheating, but in the select-ivity of his evidence. The drawings on which his case was based repre-sented relatively late stages in the development of embryos; these drawings showed the full layout of the basic vertebrate body-plan. Had he started at the logical place, the zygote, he would have realized that different classes of egg differ greatly in yolk content, size and shape, cleavage patterns, blastula, and in the organization which prepares them for gastrulation. Haeckel's series begins at the point when these diverse early stages converge, just before organ formation. This seems, for reasons unknown, to be the only tolerable intermediate stage. Thereafter, divergence again occurs into the diverse adult types. Technically Haeckel 'cheated' as much by beginning at the common intermediate stage as by filling in details from his imagination.

Several structures in the embryonic development of humans and other 'higher' vertebrates were pointed out and seized on by Haeckel and his followers as evidence for evolution; by so doing they created evolutionary legend that still persists in many textbooks. For example, 'gill-slits' have been cited as evidence that a human, about a month after conception, recapitulates the fish stage of evolution. These 'slits' are actually pairs of pouches with furrows between them which develop in the neck region of all vertebrate embryos but never, in reptiles, birds or mammals, resemble the gill-slits of an adult fish. Instead they herald parts of the lower jaw, ear and neck: these include the important thyroid and parathyroid glands. Such is embryonic engineering that the pouches are an essential stage in the guidance and supply of blood vessels to the forepart of the body. In fish these vessels divide, forming a network which extends into the gills; but in other vertebrates neither pouches nor furrows ever assume the character or function of gills. In short, they are a feature of developmental design, not evolution.

'Gill-slits' are but one example of a group of organs which evolutionists identified as 'vestigial' – vestigial organs were thought to have been functional in the history of their possessors but were no longer useful. Victorian scholarship catalogued about 180 such structures (mostly muscles) in the human body. All except about six, however, are now known to be useful, even essential. Tonsils, the pineal gland, vermiform appendix and dozens of other structures have simply been reinterpreted, though the concept of 'vestigial' organs remains in scientific – even more in popular – evolutionary literature.

Loss of organs or their function may certainly occur. Examples are found in fish or amphibians, isolated in dark caves, which have underdeveloped eyes and cannot see, and in wingless insects which predominate on certain islands while similar winged forms occur on neighbouring mainlands. But winglessness represents a loss, not a gainful step in evolution.

The issue is one of engineering concept and design. As usual, the question is whether these plans represent the broad and indivisible integrity of an idea, or whether they represent an environmentally honed hotch-potch of accidents. Is the development of zygote into an organism the product of a series of mutations – adulthood a genetic accident? Or is plan and purpose nowhere more evident in biology than in the study of development?

During embryonic development, structures, which each have their own function, appear for varying lengths of time. However brief, each plays a vital part in the drama of unfolding life. For example, the notochord (a cartilaginous rod resembling a backbone) is absent in most adult vertebrates but if it is removed from an embryo no central nervous system will develop. Structures in embryos are not atavistic vestiges: they conform to an essential economy of design.

Animals are similar in basic structure, being made of cells; they are also similar in function, as nutrition, respiration, reproduction etc. And, like the block of wood we considered earlier, early stages may resemble each other, before the divergence into the unique end-product becomes apparent. Similarity need have nothing to do with ancestry. For example, the embryos of man, cat, hen and snake are alike in having their heart, main arteries and neck region develop on the same plan as a fish. Why should a common embryological sequence, where there is one, which leads to various adult forms, not represent the most efficient engineering route to the structures which are common to all of them?

In a seminal work which helped lay the foundation for modern evolutionary thought about ontogeny and phylogeny Sir Gavin de Beer noted that '. . . the theory of recapitulation has had a great and, while it lasted, regrettable influence on the progress of embryology'.[1] It is 'fallacious'. Science has outgrown Haeckel's law.

How Would You Have Done it?

It is presumptuous to offer a blanket explanation of the global evolution of animate nature over an alleged few billion years when the phenomenon of development (ontogeny) is not well understood. Nevertheless, intense research is focused on the subject and there is no reason to suppose it will not be fully elucidated in terms of physics and chemistry. Just as robots make cars, so bodies make other bodies. It is my suggestion, however, that as these studies approach the climax of their revelation, such complexity of control will have emerged that evolution, due to natural forces, will seem insufficient to account for it. Just as the robots needed intelligent makers

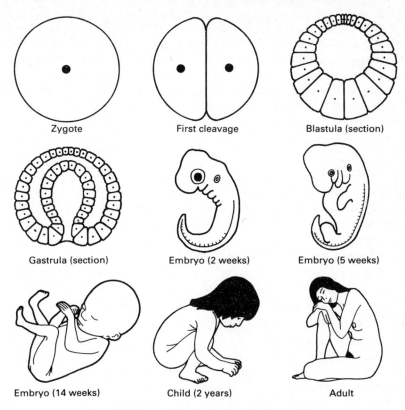

Figure 13. *Stages in the development of a human being*

and programmers, bodies will be seen to have arisen as the product of intelligence.

The literature surrounding developmental biology is now so large and growing so fast that it might seem hard to know where to start. Using the creationist trick, however, we can ask ourselves how an engineer might design a self-reproducing machine and enquire if nature has preceded him in his logic.

It would be difficult, perhaps impossible, to reproduce one multi-cellular form direct from another. A brilliant idea would be to reduce the parent to a single cell and then, from this single cell, build up a new adult. This is what happens in nature. The parent is reduced to an egg or a sperm; these fuse and the product, a zygote, follows a complicated pathway, developing into the next adult generation.

To duplicate a complicated pattern, an architect or engineer needs a blue-print. For a dynamic process such as motor-car manufacture or maintenance, or the production of a newspaper, an equivalent to a blueprint is a computer programme. This can be transcribed in whole or in part and sent round the world; whenever it appears, the process – whatever it is – can be carried out exactly. Such a programme contains all the instructions needed, carefully and economically packed in the right order, perhaps on a tape or a 'floppy disc' the size of a saucer. As we have seen, the equivalent in biological terms is the DNA of the cell, which contains all the instructions for replicating the cell. It also contains far more – all the instructions for replicating not just the cell but the whole body to which it belongs. Wasteful? Uneconomic? Not really. Does the editor of a newspaper print only the information that each individual reader requires? No, he prints the lot and lets each reader select what he or she wants to read; the extra expense is justified many times over in the economy of bulk production.

Hundreds if not thousands of chemical reactions can go on in a living cell. These must be coordinated and regulated. How can a cell choose which sections of the whole instruction concern it especially? After all, neither during the grand process of development nor in the humdrum business of maintenance will a cell need to make every type of protein or other material, simultaneously or in the same quantities. If you have recorded a song, you can use the tapecounter to find it again. Likewise, computers act on a sequential set of instructions, as contained in a programme, by using line numbers to find their way about. How does the cell switch its DNA from one process (one subroutine, in computer terms) to another as the need arises? This aspect of genetic control has exercised the minds of some of the finest twentieth-century biologists.

In bacteria, for example, Jacob and Monod demonstrated a control system that operates by switching off 'repressor' molecules, i.e., unmasking DNA at the correct 'line number' to read off the correct (polypetide) subroutines.[2] With eukaryotes Britten and Davidson have tentatively suggested that 'sensor genes' react to an incoming stimulus and cause the production of RNA. This, in turn, activates a 'producer gene', m-RNA is synthesized and the required protein eventually assembled at a ribosome.[3] Many DNA base sequences may thus be involved not in protein or RNA production but in control over that production – in switching the right sequences on or off at the right time. These researchers seem to have discovered at least part of a biological index.

A Book Needs an Index

Information, however detailed or complete, is useless if it is inaccessible. A large data bank, like a library or disc storage, needs an index. Only in this way can relevant information be extracted precisely when it is wanted; an index, unless it precisely interlocks with the system which it serves, is worthless. Indeed, it must be created as an integral part of that system.

DNA, the chromosomal book of life, is a data bank which is subject to continual demands for information. Access is obtained through a dynamic index. This means the DNA and the index reciprocate; information is cycled round a feedback loop. The index triggers the production of materials by DNA; the presence of these materials, for example hormones, can affect a new line number; synthesis of another protein occurs, and so on. Both the quality (called specificity) and the quantity has to be controlled. The analogy with an indexed computer programme is clear. One subroutine ends by directing the computer to a new line number: here it either repeats or engages a new 'block' of information, which is expressed as a new phase in the programme. If this programme can switch endlessly between subroutines, yet is sensitive to instructions from outside, the analogy with biological computation is closer still.

It is a new perspective that identifies what is not DNA – the bulk of the body such as cytoplasm, membranes etc. – as a biological index. On a higher level, systems, such as the nervous, hormonal or circulatory systems, are also identified as indices: so is the whole body and the mind, although interactions with DNA are always mediated at molecular level.

Take, for example, the egg – a single cell from which all multicellular organisms are built up. The animal pole of an egg is the point on the surface nearest the nucleus, and the vegetal pole is furthest from it. The yolk, a mixture of protein and fat, may be concentrated towards the lower end of the egg; in other words, a yolk gradient arises when the egg is formed in its ovary. Polarity can also occur due to some environmental influence – in the case of the seaweed, *Fucus*, it is light which causes a calcium gradient. Whichever way, it is thought that such an asymmetrical arrangement triggers, after fertilization, the development patterns leading to an adult form. In other words, the arrangement acts as an index. People start as fertilized eggs. Of course, all eggs need mothers; and all mothers come from eggs. Which came first? Similarly, if DNA is the book of life, non-DNA is its index. Which came first?

As the zygote develops, parts of the body appear. They are usually well formed. It is the same at cellular level and the biochemist suggests that down among the molecules, an invisible scheme is no less defined. At this point the creationist offers a second shift in perspective. The index code and its reciprocal DNA could be radically conceived, not in terms of molecules or even atoms, but rather of electrical polarity disposed in several ways about the cell membrane, cytoplasm and nucleus. The shift is therefore from a chemical to an electro-physiological perspective.

In this way the body is reduced to binary digits disposed in the form of electrical polarities, i.e., (+) and (−) charges. The charges exist in the form of ions or polar bonds, where electrons have been pulled or pushed towards one end of the bond, creating an electrical tension. Chemists call alcohol, amino or carboxyl groups 'functional' groups because they contain highly polarized bonds, bonds ready to react.

At molecular level a body is a labyrinthine network of chemical communcations. These communications are based on precise, three-dimensional distributions of charge at reactive sites. Such sites form, for example, the basis of enzyme activity. In some cases there is interaction between DNA and non-DNA chemicals. DNA is an information-storage system, a file: the non-DNA chemicals address specific locations in the file from which information is retrieved. A subsequent print-out is obtained in the form of correct biochemicals to satisfy the organism's changing needs. The non-DNA chemicals thus index the DNA.

The body is reduced to fundamental units, the presence or absence in space of an electrical charge. Given the right equipment, it ought to be possible to map the structure and monitor the dynamic condition of this invisible electrical 'ghost'.

It is clear that this subatomic, electronic aspect of being could be viewed as a sort of stereo-computer system. The binary digits of this biologic machine, as it computes for life, are (+) and (−) charges. According to this stereo-computer model the architecture of a protein or nucleic acid is a framework which supports, protects and presents a three-dimensional electrical shape, made of ions or functional groups and exact down to the last electron, so that it can interlock with another exact electrical shape and promote a required reaction.

In seeds it is found that change in a single gene brings change to the overall electrical pattern of the chromosome. There seems to be a close relationship between the genetic constitution and the electrical pattern. Likewise proteins, including the histones which are bound with DNA to make a chromosome, are charged molecules sensitive to their electrical environment. Differing electrical 'ambience' in different cells is calculated to 'break into' different subroutines of the genetic code. Does this 'ambient' index inform the complex process of cell development? D. MacKenzie thinks so:

The epigenetic input − that is, telling a cell where and who it is and therefore which genes it would be appropriate to use − is after all the crux of developmental interactions, the dynamic process through which differentiation and morphogenesis are enacted. Many of these instructions, it now appears, ride along on currents of electricity.[4]

Shaping up

This can only be part of the story. DNA certainly needs an index; morphogenesis, however, is a kaleidoscope of targeted effects which reach from the molecular up to the visible architecture of life. If the chain of events flowing from DNA stops at basic protein structure, how can the different shapes of cells, tissues, organs and organisms be explained by DNA alone? How does it all shape up?

The creationist takes his first standpoint among the archetypes and phyla discussed earlier (p. 37). A phylum, you may recall, is a group of species which share a common body-plan, usually with one or more unique organs. Examples are the molluscs, chordates, annelids, echinoderms etc. No new phyla have appeared in the fossil record for 350 million years. Then, as now, they appear distinct. There is, and always has been, an absence of transitional sequences between phyla: they are not artificial groupings which help biologists to classify life, but distinct, natural entities.

D'Arcy Thompson, a brilliant zoologist, mathematician and classicist, showed in his theory of transformation that a regular mathematical relationship exists between certain forms: in other words, you can derive different forms from a basic shape or archetype.[5] It is as if a shape on a rubber sheet were transformed by pulling the sheet in one direction or another.

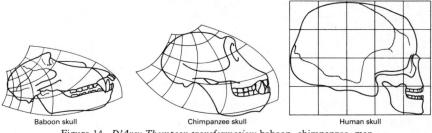

Baboon skull Chimpanzee skull Human skull

Figure 14. *D'Arcy Thompson transformation:* baboon, chimpanzee, man

Thompson did not explain the origin of the archetypes. Given that both today and in the fossil record, discontinuity seems to be the rule, how did one archetype transform into another? It is here that we address the nub of the problem. For the evolutionist, transformism (i.e. macro-evolution) occurred. For the creationist, just as there exist genetic constraints (p. 67), there are limits beyond which a design structure fails to work. Phyla represent 'taxonomic gaps', unbridgeable structural differences between which intermediate structures are unfeasible. How, as a mechanical example, could a series of viable engines exist between an oscillating and a rotary engine?

Stephen Gould suggests that 'punctuated equilibrium' leaves no evidence of transformism in the fossil record. He espouses a second idea that, he believes, might cause macro-evolution without leaving a trace. Growth, change in shape and onset of sexual maturity are three important parts of development. In normal development these are time-coupled in a set way; however, if mutation in 'rate genes' caused alteration in this time-coupling, one factor or the other would be accelerated or retarded. Although Thompson's illustrations were of adult creatures, actual transformations are thought to occur by differential growth rate in different parts of the embryo during development. In this way, a minimum of genetic mutation that caused rate changes early in development could lead, by the time adulthood was reached, to a large variation in shape. Thus, similar embryos might bridge the 'taxonomic gap' in their adult form. Is this how archetypes invisibly evolved?

For the creationist it is not. For example, a salamander called the Mexican axolotl remains a tadpole, though it is able to reproduce, because it cannot make a certain hormone; it has not bridged the 'taxonomic gap'. The idea that 'rate genes' can cause macro-evolution is subtle, invisible and plausible but no evidence exists that it is true. Equally, then, just as an artist organizes molecules of paint or a musician the vibrant energy of sound (p. 30), so the creationist prefers that biological archetypes – at every level – are the product of Ideas. Archetypes include not only adult phenotypes but the developmental process itself, from its roots in the sex cells and zygote. Just as mind's interaction with body is electro-chemical, occurring within the cellular circuits of the brain, it is suggested that an idea preceded and generated the electro-physiology, with its DNA correlate, of each archetypal form. That such conceptions should be realized in the materials available from the parent planet is natural. Just as we exploit an instrument's capacity for sound, so life is programmed to exploit – to the limit – the chemical nature of its parts.

Despite some plasticity of body-plan it is usually obvious to which of about thirty phyla a species belongs, and the common features of a phylum can be represented as an ideal pattern, a summary structure known as the archetype. For the evolutionist this is the prototype or 'simple' pattern from which 'advanced' or 'specialized' forms have evolved. For the creationist it is the basic idea from which radiated the actual, specialized *creaturae* in the phylum. 'Primitive' archetypes are rare in the rocks; even the monoplacophoran mollusc, said to be an example of such an archetype, is a robust and successful *creatura*. Whence did *it* spring?

An enormous diversity of forms could have evolved, or been created, from the relatively small number of arche-animals or 'ideals'. The distribution of archetypes would, according to the creationist, be ecological rather than evolutionary. Life's biology is conceived as a whole. A mosaic of subroutines, generated within the broad outlines of a phylum or a class, would be designed to fit each *kind* of organism directly to its surroundings. For example, the vertebrate programme has been modified to fit creatures directly for life in water, on land, in air and in specific terrain, vegetation and climate.

Developmental Biology

Development of a new form is clearly targeted towards procreative adulthood. A target is the product of forethought. If you were the designer, how would you specify the self-assembly of a working machine that must work at every stage of its development? The morphogenetic (shape-making) writ runs beyond the mere aggregation of cells. For example, your arms and legs contain identical cell types (muscle cells, connective tissue cells, bone cells etc.) with identical proteins and DNA. So the differences can hardly be ascribed just to DNA. There must be some pattern-making feature which differentiates between an arm or a leg – even between a left leg and a right one.

Here is an example of the kind of complexity we are dealing with. In the development of vertebrate eyes, fine filaments spread out from a million or so ganglion cells in the retina, and each 'homes' along the optic tract to a precise location on the visual cortex of the brain. This location corresponds exactly to its image point on the retina. Now if, at a certain stage in the development of an amphibian embryo, the eye is experimentally inverted, the dendritic filaments adjust, home correctly, and the animal sees normally. If the eye is inverted at a slightly later stage, the filaments cross over and home in such a way as to produce an inverted image. How are these filaments guided? No, it has been shown that even if the cells of the retina and cortex are separated from each other and placed in tissue culture they are still able to associate in their particular patterns.[6]

Several suggestions have been put forward to explain 'shaping up' of this kind. For example, some biologists believe that DNA and protein synthesis are sufficient to explain the development of form. If we can understand how such synthesis is controlled we will have solved the riddle. It is, of course, true that the make-up of chemicals predisposes them to assume certain structures. Crystals grow into predictable shapes. Lipids spontaneously create the bi-layers found in cell membrane and fragments of the protein tubulin link to form rod-like microtubules of the kind that make up the cell's skeletal support.

Cell membranes exhibit, due to their physical and electrical constitution, various degrees of 'stickiness'. It is believed that different cell types are 'sorted out' according to their 'stickiness'; adhesions between unlike cells are exchanged for those between like, and more adhesive cells move towards the centre of a group. In amphibian gastrulation, for example, it is perhaps a combination of cell-to-cell adhesion forces that transmits signals which cause many cells to migrate towards an expanding hole in the gastrula (the blastopore), roll over its lip and disappear inside. Certain tissues can spread over others but not vice versa; the cells from outside can, in this case, spread out over the inner surface of the gastrula. There seems to be a hierarchy of 'spreading' incorporated into an overall programme of morphogenesis. Is it not possible to conceive that stickiness' might be programmed to vary at different stages in development, generating different molecular aggregations?

The surface of a cell is studded with protein receptors. Each greets only one sort of visitor, such as a certain hormone, and their embrace signals to the production lines inside the cell to swing into the required routine. In some cases, such as hormone-secreting cells, the structure and function of a cell is designed to service a dominant routine.

In addition to receptors all cells carry exactly defined chains of sugars anchored into position on the external surface of the membrane by a protein or a lipid tail. They are like antennae or identification tags by which a cell can know, and be known by, its neighbours. Tags are important for the operation of the body's immune system, which fights to keep you well. The system will attack any cell with an unidentified label. Perhaps a related system of polysaccharides, mounted on a protein base, plays its part in conveying strategic information during morphogenesis. If so, it would constitute another complex part of the index code.

Life is *programmed* to exploit the chemical nature of its parts: it starts from a more or less spherical egg and from this there develops an animal which is anything but spherical. From a zygote develop many sorts of cell; and, as a Cambridge-based research team lead by Sydney Brenner has found, in a tiny nematode worm even a *final* cell division can produce one nerve and one muscle cell. Can we account for this by a theory which confines itself to chemical statements, such as that DNA controls the synthesis of certain proteins? Can this alone account for why you are human-shaped and a cat is cat-shaped? Apart from mutations in regulators called 'rate-genes' it has been suggested that 'fields' might influence the development of biological shape. Their interaction would be strictly ordered through to an adult conclusion. Although their nature is disputed, the shape of inanimate matter is certainly influenced by such variations as sound, electromagnetic radiation, atomic bonds or the electron. The German physicist Ernst Chladni showed that inorganic matter vibrated with sound assumed 'organic' shapes. Different notes on a violin vibrated sand on steel discs into 'natural' spiral, circular, hexagonal and other geometries. Rupert Sheldrake even suggested that an extension of what we call matter, i.e. matter and material energies such as nuclear, gravitational or electromagnetic forces, might include 'morphogenetic fields' which organize the shape of all objects, whether animate or inanimate.[7] Whether or not his exotic theory is right, it is a fact that shapes are the visible expression of invisible, vibratory energy. Form and movement, like light and colour, are essentially a matter of wave patterns. These patterns relate harmonic (p. 30), electrical, chemical and electro-physiological principles to the shape of visible objects.

Like a magnetic field, this 'shape matrix' could be intangible, invisible, tasteless, soundless, wireless and odourless, but still present, active and capable of being altered by other electrical stimuli. In 1935 H.S. Burr, Professor of Neuroanatomy at Yale University and Dr F.S.C. Northrop established that all living matter, from an amoeba to a rhino, from a seed to a human being is controlled by electrical fields.[8] Burr, who experimented with salamanders, claimed that these commonplace amphibians can be shown to possess an electrical field complete with a positive and negative pole, arranged along the longitudinal axis of the body. According to L. Watson, Burr traced the development of the field back through the growth of the embryo, finding to his astonishment that it existed even in the unfertilized egg.[9] He marked with blue dye the pole of the egg where there was a noticeable drop in voltage, and noted that the head of the salamander always grew opposite to that point. The embryo cells were, in fact, arranging themselves according to the pattern of an electrical field that was there before the individual came into existence.

Burr's work has not formed part of the trends and fashions profitably centred in biochemistry and molecular biology, which have shaped biology over the last forty years or so. As a result, no underlying theory of electrophysiology has been developed. All cells, however, possess a significant set of electrical properties and interesting patterns of electrical and chemical energy are fundamental to the problem of organization in biology.

Cell voltages are generally too low to be picked up by conventional micro-electrodes, and their detection awaited the necessary equipment. Burr used a vacuum-tube voltmeter which required virtually no current for operation. Now a vibrating probe is used. This is a platinum electrode which, vibrated between two points outside the organism, measures the voltage (in millionths of a volt) between them.

The modern study of electro-physiology is practically restricted to the study of nerves and muscles. For example, nerve outgrowth can respond to small electric fields, some tens of millivolts per millimetre. Nerve and muscle cells seem to like being in an electrical field. Nerve-cell growth cones head for the negative pole when a field of as little as 300 microvolts is placed across their width. They line up parallel to the field whereas a muscle cell will line up in a perpendicular way. This is, in fact, their orientation during development.

Burr suggests that man and other life-forms are ordered and controlled by 'electrodynamic fields' that can be measured and mapped with precision. No doubt, the relativistic concept of a wireless anatomy requires as different an approach to body (and, perhaps, mind) as that of classical Newtonian to quantum physics. Suppose that a 'field' is composed of the entire electrical ambience of a cell. This includes its pH, its transmembrane voltages, the passage or accumulation of any charged particles and the presence of reactive chemical bonds. Suppose also that the shape of a biological 'compartment' (an organ or part of an organ) has its own electro-physiological pattern which, of course, is predetermined at genotypic and phenotypic levels for every stage of development including the 'final, target stage – procreative adulthood. How might these fields be generated from the first?

A Developmental Archetype

Along an inevitable path, with every step engraved since the dawn of human history, a zygote turns into a human being in about fifty-six hierarchical, algorithmic steps. The 'docile', generalized single cell is progressively refined into many millions of specialized cells. Determination soon sets in, to the extent that, at the gastrula stage, fate-maps for various cells can be worked out using marker dyes. A succession of stages, each with its centre of influence, progresses outward from the zygote. The hierarchical data structure involved in this developmental computation is a tree; you might say, a tree of life. The levels of the data structure, represented by 'nodes', are both spatial and temporal. (figs. 15a and b).

Spatial integration from node to node at the same stage (e.g., C_1 and C_2 at stage^2), is by induction. That is, one embryonic cell or tissue will influence

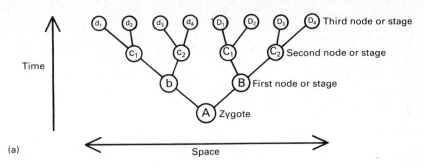

Figure 15. (a) Stereocomputation; the 'tree' is a data structure which can be used to represent the development of an organism from *zygote to adult*.
(b) This 'tree' illustrates *compartmentalization*

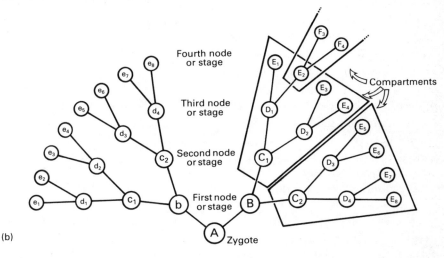

others near it. Such induction might be 'positive' or 'negative'. If positive, it would cause its neighbour to change into a new kind of cell or tissue, triggering one or a whole chain of responses. If negative, 'compartments' or segments will arise and these will, in turn, become progressively determined into part and whole organs. Two compartments starting from neighbouring centres will develop to interlock in the totality of the body, emerging like pieces in a changing 3-D jigsaw puzzle.

Inductive signals between cells or tissues arise due to position and to cytoplasmic or cell-surface receptors: they are, it is emphasized, as much code as the DNA with which they interact. Spatial signals are, it is proposed, progressively 'compartmentalized'. In 1891 Hans Driesch separated the cells of sea urchins after the first and second divisions, i.e.,

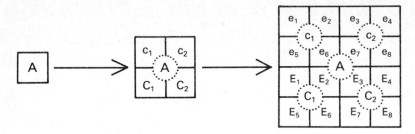

Figure 16. A schematic diagram of *compartmentalization* with centres of influence derived from the previous node. In its course of development the 'parent' zygote 'A' will become, through 'daughter' generations, different kinds of cell.

For simplicity, only stages 2 and 4 have been shown. The dashed circle signifies that 'parent' cells disappear into 'daughter' generations. They may or may not retain the cell type. For example, 'daughters' of 'A' do not remain fertilized eggs but change into all the sorts of body cells

when there were two or four cells. Each cell produced a complete embryo. Separation of cells at later stages led to abortion. In plants, the position of root and shoot is usually established in the first division of a zygote and the polarity passed thence to each of the cells in a multicellular plant. After a single division half the cells are fated one way and half another (fig. 15). At successive stages new centres of influence arise (figs. 15b and 16) and generate, by virtue of their position and composition, new electrophysiological feedback for the DNA.

Temporal control would not only involve neighbouring inductors: the development of the organism as a whole would have to be integrated. Different centres of development would, at each stage, have to be kept 'in step'. In practice, this appears to be the job of pacemaker proteins – particular hormones secreted from glands and active on target organs. Their effects can be general or localized, and the same hormone can trigger different DNA subroutines and generate different shapes in different parts of the body. An obvious example is the sex hormones involved in generating a male or female embryo. Whatever the details of the biochemistry, it would be precoded and form an integral temporal aspect of the index code.

It is easy to understand that the switching between the non-DNA index code and DNA demands a high grade of pre-programming. In development, no stage by itself has any value in terms of natural selection: instead, it represents an intermediate condition (like enzymes on a metabolic pathway) in a route towards a goal. How, therefore, did the overall pre-programming, initially required to control the development of an organism, arise? The creationist suggests that development is an archetypal theme, one required to reconstruct any multicellular organism from a zygote. Just as an index is created with the information it serves to render accessible, so an inextricable unity compounds a body with its productive DNA. This unity must have been present from the start.

The Long and Short of it

If the sensitive interdependence between DNA and the index code is as precise and pre-programmed as this chapter suggests, this leads to several conclusions. Firstly, because the index (in the form of egg and sperm) trips the DNA code into its developmental sequences, it might be termed the initiator. It commands a role akin to that of a conductor over the score. It 'leads the dance'. This relegates genes from their current star role in molecular biology. Both codes are equally important.

Secondly, because sex cells derive from a multicellular source, the cell cannot be regarded as the basic unit of multicellular life. That would represent a form of atomism which lacks a morphological explanation. Although the zygote is a convenient base from which to redevelop a multicellular form, the origin of life needs to be thought of in terms of chickens rather than eggs.

Not much has been written about the evolution of morphogenesis or metamorphosis. No one, however, suggests that children preceded adults and, by a series of mutations, developed into them. Experiments indicate that, even in phages (a kind of virus), the assembly of components such as head and tail are under separate genetic control, i.e., they are assembled separately and then 'click' together spontaneously. In other words, the origin of life at every level needs to be thought of in terms of the origin of *whole* organisms rather than of self-replicating DNA molecules which carry hereditary information but lack genetic machinery and the context of life in which to express it. The origin of DNA (p. 146) is problematical enough: a cyberneticist would agree that its precise integration with a second code tries the formula of chance mutations past breaking-point.

These facts, to the creationist, bespeak an initial 'composition'. At creation the *creatura* was programmed to reproduce itself in the way we have discussed. In the discipline of developmental biology, creationist and mechanist concur except on just one point — a work of art, a machine or a body which can reproduce itself cannot first make itself. The clear targets, programmes and strategies of development underline the fact that it must be purpose-built.

Notes

1 De Beer, G., *Embryos and Ancestors*, Oxford University Press, 3rd edn, 1958, pp. 13, 168.

2 Monod, J., *Chance and Necessity*, Collins, 1970, pp. 75–7.

3 Britten, R.J., and Davidson, E.H., 'Gene Regulation for Higher Cells – A Theory', *Science*, vol. 165, 25 July 1969, pp. 349–57.

4 MacKenzie, D., 'The Electricity That Shapes Our Ends', *New Scientist*, 26 January 1982, p. 220.

5 Thompson, D'Arcy, *On Growth and Form*, Cambridge University Press, 1917.

6 Gaze R.M., 'The Problem of Specificity in Nerve Connections' in *The Specificity of Embryological Interactions*, ed. D. Garrod, Chapman & Hall, London, 1978.

7 Sheldrake, R., *A New Science of Life*, Blond and Briggs, 1981.

8 Burr, H.S., *Blueprint for Immortality*, Spearman, 1972.

9 Watson, L., *The Romeo Error*, Coronet, 1976, p. 137.

9 Genesis

How can one do without the other? A computer programme or an encyclo-paedia needs an index: in order to perform its task DNA needs a body to index it. The reciprocal nature of sexuality has also been noted: the last couple of chapters have sought to integrate rather than isolate different aspects of reproduction, i.e., the origination of fresh bodies. Now it is time to examine how, from isolated and uncoded parts, the evolutionist believes the first body may have evolved.

How, in the chemical sense, did life begin? We saw earlier (p. 49) that, for anyone who needs to eliminate a Creator from his philosophy, chemical evolution is a necessity. The life's work of many scientists, supported by millions of dollars of grants-in-aid, has been devoted to demonstrating the 'fact' of chemical evolution. Certainly every living thing on earth uses the same amino acids, the same nucleic acids, fats, sugars and salts. The simplest and the most complex organisms use the same language to trans-late DNA into protein, even some of the same enzyme systems and compo-nent parts. But this indicates chemical evolution no more than it indicates intelligent, economical creation from proven chemical building units and information systems. For example, every motor engine employs similar metals and fossil fuels.

Some, but not all, creationists query the large time-scales required by the evolutionist. Here, for the sake of argument, the orthodox estimates of billions of years are accepted. In fact, not time but design is the crux of the issue.

Professor Ponnamperuma, director of NASA's chemical evolution branch at Ames Research Center, California, has noted that the lead-up to life ('prebiotic synthesis') appears as a two-part problem. Firstly, how did the small molecules necessary for life come into existence? Secondly, how were they combined, under conditions existing at the time, into the polymers or chains of molecules which are the precursors of nucleic acids and proteins? Simple organic molecules have been detected in space.

These include formaldehyde (an important precursor of sugars), hydrogen cyanide and ammonia (for amino acids and bases). DNA, you may remember, is made up from sugars and bases (with a phosphate group); and protein from amino acids. Could molecules of this kind have formed spontaneously on the early earth?

The Wrong Atmosphere

We do not know how the solar system, including earth, was formed, and we are hazy about the conditions that existed on earth and − more importantly − in the atmosphere during the early days. We do know, however, that many of the compounds used in living material are chemically unstable in an oxidizing atmosphere such as the present one. In the presence of oxygen they tend to decompose rather than build up. So Oparin, father of the modern alchemy that seeks to explain life's origins, was led to propose an absence of oxygen in the primeval atmosphere. *Without this assumption the whole evolutionary scenario fails*, for even the simple organic compounds − the smallest bricks of living material − would have crumbled as soon as they formed if oxygen were present.

The original atmosphere, he said, was a 'reducing' one, without free oxygen. It was perhaps made up of methane, ammonia, nitrogen, water and hydrogen. Some geologists have argued further that, as with the outer planets today, a primary atmosphere of hydrogen, helium, methane, ammonia and carbon dioxide enveloped the earth, and that the lightest gases − hydrogen and helium − were subsequently lost to space. But if this were true, the present atmosphere would almost certainly have retained larger amounts of heavy inert gases, such as krypton and xenon, than the mere traces we find today.

Alternatively, instead of being an original gaseous envelope, earth's atmosphere might have developed gradually as the product of volcanic emissions. Carbon dioxide, steam, nitrogen, oxides of sulphur and, with other gases, just a little chlorine would have evolved. Much of this mixture would have dissolved in sea water which is thought to have remained relatively unchanged for 2.5 billion years. There is, anyway, little likelihood of high concentrations of methane, ammonia or any other possible precursor of organic substances (even the simplest ones) arising from volcanic emissions. Perhaps earth picked up ammonia in passing through a cloud of gases in space. But that is a long shot indeed.

This is all very speculative. Does the record of the rocks tell us anything more substantial about the early atmosphere? At best it is equivocal. There are traces of iron sulphite (pyrites) and uraninite (an oxide of uranium) in some of the earliest sedimentary rocks; some would say these could not have persisted had there been free oxygen in the atmosphere at the time they were laid down. Other geologists say no — these same materials are being laid down today, when there is plenty of free oxygen.

'Redbeds' — thick deposits of sand stained with oxidized (rusty) iron — occur only in rocks less than 2.2. billion years old: could this be evidence that the atmosphere was relatively oxygen-free before that date? Perhaps; this correlates nicely with the presence of stromatolites — hard calcareous structures believed to have been formed by blue-green algae — in rocks of a slightly earlier age. Did the algae provide the oxygen that rusted the iron to make the redbeds? Again — perhaps, but the case is far from complete. Because much of the oxygen that is present in the atmosphere today is derived from photosynthesis (the chemical process by which green plants split water into its component elements), it is tempting to get into a circular argument and assume that oxygen was rare or absent before life began.

Clearly, oxygen from photosynthesis was not available if plants were not present; but to assume that absence of life means absence of oxygen is wrong. R.T. Brinkmann has calculated that an important abiotic process — the breaking down or photolysis of water vapour by intense ultra-violet radiation — could have generated considerable amounts of free oxygen in the earth's early atmosphere, without life being present at all. Other evidence exists that free oxygen was probably present at that time. Indeed, at a recent conference it was generally agreed that earlier views, involving highly reducing conditions, were now giving way to geochemical evidence for a nearly neutral atmosphere.[1]

So much for the 'reducing atmosphere'. The fact that the Oparin model needs such an atmosphere does not mean that it ever existed; independent evidence suggests that oxygen might always have been present. But suppose for a moment that the atmosphere was right for biosynthesis and that simple, stable organic compounds could form and persist — what then? Could they have built up into the larger molecules that life seems to need?

Flies in the Soup

Proponents of the Oparin model suggest that molecules concentrated into a 'soup' — a watery solution and suspension in which the further synthesis of larger molecules took place. The concept of 'primordial soup' gained strength from the laboratory experiments of Stanley Miller and others, in the United States, who passed an electrical discharge through a mixture of methane, ammonia, water and hydrogen and found they had synthesized small organic molecules, including amino acids, in the resulting brew. Amino acids have also been produced by irradiation (glycine and alanine), electron bombardment and thermal synthesis (at about 1000°C.). Methane, ammonia and water were used to simulate a primitive atmosphere. Electron bombardment also produced hydrogen cyanide, from which amino acids and bases may be obtained. But if hydrogen cyanide had been present on the earth in early times, the iron-cyanide complex, Prussian blue, should be present as a mineral: yet it does not seem to occur at all.

There are several reasons why these experiments are not relevant to events on the primordial earth.

(a) The reaction products in these experiments are usually isolated as they form by a 'cold-trap' which removes them from the destructive influence of the electric-discharge zone. This could not have happened in the natural environment which the experiments set out to mimic.

(b) Isolation of the amino acids would have the effect of enhancing yields well beyond any natural level. Bearing in mind the enormous volume of water in the seas, it is difficult to believe that the concentrations of organic material required for polymers could have occurred. D.E. Hull has calculated the likely concentration of the simplest amino acid, glycine, in a brew of the kind that might have developed.[2] It was thin soup indeed – the equivalent of 0.2. mg of organic material in a swimming pool full of water – and this is the concentration of one of the simplest and most probable compounds. Sugar yield would work out far lower. Hull concludes that the physical chemist, guided by the proven principles of chemical thermodynamics and kinetics, cannot offer any encouragement to the biochemist who needs an ocean full of organic compounds to form even lifeless coacervates (gel-like globs). Moreover, under the conditions of the proposed synthesis, the rate of decomposition exceeds the rate of production by a factor of 10^4. It has been suggested that some surfaces (e.g. rocks and clays) might absorb the reactants and so catalyse the process. But this only increases the rate at which the same poor yields are reached; it does not necessarily improve the yields themselves.

(c) Every amino acid found in proteins except one (glycine) exists in two forms which are mirror images of each other (like left and right hands). With very few exceptions only the so-called L-form (left-handed) occurs in living proteins. Where D-forms (right-handed) do occur, it is by deliberate synthesis by enzymes. Otherwise, D-forms destroy enzyme activity: they are edited out of the system. Amino acids, generated by discharges of electrical energy through 'soups', contain both forms of amino acids. A moment's thought will show that this is a very unlikely starting point for a

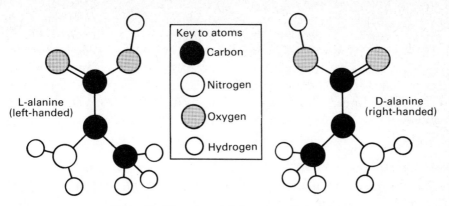

Figure 17. *Mirror images of the amino acid alanine*

system of life-chemistry that uses only one form of amino acid. Since the two forms are almost identical in their chemistry, how could one (the L-form) have been selected while the alternative (D-form) was rejected? Why didn't an alternative kind of life based on D-form or mixed amino acids arise? Had experiments produced only one of the two, or even L-form amino acids in preponderance, biochemists would have been convinced that this is how it might have started; but present experiments are *not* a simple model of a life-forming process, and only the very gullible would accept them as such. Incidentally, at least three key amino acids have not yet been found among the products of these experiments.

(d) There have been many successful attempts to produce biologically significant chemicals by natural means – methods that could have appeared naturally in an early environment. Ribose, deoxyribose and other sugars can be built up by irradiating methane, ammonia and water, or from formaldehyde. Adenine, a base unit of nucleic acids, can be produced from hydrogen cyanide, and nucleotides may be synthesizable by irradiating adenine, ribose and phosphoric acid with UV light. But none of this simple, clever laboratory chemistry could have happened in the open-air conditions of the primitive world. There were no pure chemicals obtainable from suppliers, no isolating techniques to keep reactive end-products apart. Sugars and amino acids would have interreacted quickly to form non-biological products, not waited about to be built into protein polymers or starches; fatty acids would almost certainly have disappeared into non-biological reactions as soon as they were formed.

(e) If lightning had struck, rather than the comparatively mild electric discharges used by Miller, organic chemicals present could not have survived.

(f) In the present earth's atmosphere, a layer of ozone (a form of oxygen) at great height protects living matter from potentially lethal doses of UV radiation. If there was little oxygen in the primitive atmosphere then there was less ozone too. It is hard to say precisely what the results could have been, but there would have been little chance of survival for any forms of life in the upper layers of a 'prebiotic soup', and none for any that attempted to emerge from its watery protection.

Building up Chains

There is no doubt at all that complex organic chemicals can be built up from simple inorganic chemicals, in a variety of ways, and nature's own laboratories might, from time to time, have produced some of them spontaneously − perhaps even in soups of some kind. But we are still a long way from showing how the organic units − sugars and amino acids, for example − could have built up spontaneously into the polymers or chain molecules that form the substance of living materials. As we have seen, the soup would have been very, very dilute. But, assuming that some of the links had been synthesized, how could they have been strung together?

Polymers are the product of condensation reactions. Energy is needed and a molecule of water is expelled for every linkage that occurs. Condensations could not occur in 'soups' of the kind we have been postulating. There the opposite process − hydrolysis − is far more likely to occur, with the absorption of water and the breakdown of the chain into its component links. To reverse this process and condense amino acids into a chain, a cell employs a complex set of molecules (see fig. 7c). This implies an extensive back-up system behind the process.

In the laboratory we can, as did S. Fox, polymerize amino acids into something resembling proteins by heating a mixture of them to 175°C.; washed and filtered, the product includes spheroids of very short-chain, protein-like compounds. Could such a process have occurred in nature? Perhaps it could on the side of a volcano, but how the amino acids got there (in pure enough state) and what happened to the products after they were formed, are relevant questions. No one could seriously suggest this as a model of how, in nature, proteins evolved from their components.

A most important component that is missing from the short-chain compounds formed in this way − indeed, from all the products of laboratories that specialize in producing organic compounds by 'natural' processes − is order. The proteinoids produced at high temperatures, for example, are essentially random chains − far removed from the highly ordered, codified proteins and protein-derivatives of even the simplest forms of life. They have potential, like stone blasted from a hillside or the components of a watch, but their randomness sets them apart from life-induced equivalents. No coding or 'intelligence' has been built into their structure, and there is nothing about them to suggest that they could build themselves up into anything more complex. They are a long way from DNA, haemoglobin, or even the simplest enzymes, which are the product of a code. And code, as we have seen, is a form of thought. Furthermore, they would have included both L- and D-forms of amino acids, which effectively excludes them as precursors of life (p. 140).

Francis Crick estimates that for a modest protein, 200 acids long, there is one chance in 10^{260} that any given specific order will occur accidentally.[3] This is useless for life which needs to print proteins like newspapers! If, by chance, one were generated, another would be required to utilize the product. Without these complementary enzymes, the single enzyme activity would not only be useless but destructive – enzymes can break as well as make. A simple cell requires at least a hundred different, specific protein molecules plus sugars, lipids, nucleic acids etc. One lonely protein in an expanse of the 'soup' would be destroyed long before it met so many biological 'friends'; and, if such an unlikely combination did float together, how was it sufficiently separated from similar molecules whose presence would have fouled up the system?

It has been suggested that colloidal minerals, such as montmorillonite clay, might have absorbed amino acids and acted as prebiotic templates for peptide (amino acid) chains. Defects in the crystal structure might serve as 'primitive genes'. It is hard to see why, in such a system, any particular pattern (such as one necessary for a life-form) should ever be selected out. Nature left to itself, with no outside influences, leans always towards randomizing rather than specifying processes. How long are these chains supposed to last in the wind and rain? Also, the crucial DNA coding is absent. How is this supposed to 'relieve' the clay? There is no evidence whatsoever, despite evolution-driven inference, that your ancestor was a block of clay or any part of it.

M. Eigen, from the Max Planck Institute in Germany, has proposed that 'guided chance' could do the trick.[4] His hypothesis assumes a rich 'soup' with huge quantities of proteins, DNA and RNA – not, as we have seen, a likely premise. He postulates hypercycles – cyclic pathways in which molecules like RNA or protein arose by chance from random beginnings and were selected as 'the fittest molecular assemblies'. This is biologically meaningless because the 'fitness' is chemical, and chemically there is no great distinction between one such assembly and another.

Eigen believes that life must originally have contained both D- and L-form amino acids and sugars before evolution found a way to eliminate the D-form amino acids and L-form sugars – though no evidence from molecular palaeo-biology supports this idea. Finally, Eigen himself does not believe that the evolutionary process directing chemical developments can ever be reconstructed in detail, though, like many another neo-Darwinian, he wants to feel that such self-regulating and self-starting processes were possible. We will need better evidence than he has provided to support his hopes of a secular explanation.

Starter Motors

'Living matter', comments leading enzymologist M. Dixon, 'is the most wonderful chemical system in the world.'[5]

Part of its 'magic' is that it consists of a complex network of chemical reactions and processes, arranged so that the product of each reaction is the starting material of the next in the chain. All such reactions are brought about by enzymes, of which there are many thousands. These are special proteins, each with the power of causing specific chemical reactions that would not occur in their absence. Dixon likens enzymes to automated machine tools, each of which performs one particular operation on a product and hands it on to the next. Some production lines join up, giving rise to a network of lines with many pathways – a network called metabolism.

Metabolism includes both breaking-down and building-up reactions, all of which need energy. In some of the synthetic (building-up) reactions, enzymes are among the end-products, and so self-sustaining. Enzymes are, essentially, ordered sequences of amino acids each derived from a precise DNA code sequence. Parts of the acid chain fold into a helix and then the whole folds into a twisted, 'higher order' structure containing an active site at which the enzyme will bind with the substance it is going to affect. These sites are normally highly specific, and will react only with particular substrates. Their specificity provides rigorous control over the chemical production line, so that transformations – for example, that of glucose into carbon dioxide, water and energy – occur only along well-defined 'metabolic pathways'. The plan or pattern for metabolic pathways is thus determined by the sequence of enzymes employed and this, in turn, is chemically coded into the cell's nucleic acid (DNA) which is, itself, built up by enzyme action.

Dixon confesses that he cannot see how such a system could ever have originated spontaneously. The main difficulty is that an enzyme system does not work at all until it is complete, or nearly so. Another problem is the question of how enzymes appear without pre-existing enzymes to make them. 'The association between enzymes and life', Dixon writes, 'is so intimate that the problem of the origin of life itself is largely that of the origin of enzymes.'[6]

A living cell is an integrated system which, let it be repeated, depends for its life both on its content of active, specific enzymes catalysing chemical sequences, and on life-support mechanisms which must spring, from the start, into simultaneous, integrated operation. A half-measure reproductive system giving semi-mitosis would not work. Enzymes 1, 3, 5 and 7 of a sequence are useless without 2, 4 and 6; any absentee blocks the system, rendering it a non-starter. Certain basic pathways are common to all living systems and must have been present since life began. The glycolytic pathway in which sugar is broken down, releasing energy, is one example; respiration and many other functions basic to life also depend upon networks of metabolic pathways, usually involving dozens of complex stages, which are recognizable throughout the plant and animal kingdoms. It is difficult to believe that the elegance and effectiveness of these pathways were achieved by rapid evolution long before the organisms themselves had diversified, and that they remained static while the plants and animals evolved around them.

Enzyme systems are doing every minute what battalions of fulltime chemists cannot. The mechanisms of their actions are only just beginning to be understood and we cannot yet manipulate them with any confidence. The idea of designing enzymes for specific purposes, then synthesizing them, is futuristic. This may come: if it does it will be the product of very concentrated thought and manipulative skill by teams of dedicated scientists. Can anyone seriously imagine that naturally occurring enzymes realized themselves, along with hundreds of specific friends, by chance? Enzymes and enzyme systems, like the genetic mechanisms whence they originate, are masterpieces of sophistication. Further research reveals ever finer details of design. Chemists engineer large molecules, like drugs and plastics, for specific jobs. The creationist view is that biomolecules — and the capacity to manufacture them — were likewise deliberately engineered.

DNA

The genetic code, embodied in DNA, underlies and informs all living systems. Codes and codification, as we have already mentioned, represent the opposite of chance. To understand life's origin it is necessary to understand the origin of the coding systems and the codes themselves that control it.

DNA is a fairly complex molecule (fig. 7a). How is it supposed to have 'arisen'? Bases in the code are of two kinds: purines (adenine and guanine) and pyrimidines (thymine or, in RNA, uracil; and cytosine). Four of these five have been synthesized using hydrogen cyanide or cyanoacetylene; we have not yet managed to make thymine. Sugar production is relatively easy in the laboratory. The third important component of the nucleotides is the phosphate group. Surely this could be easily obtained? Curiously, it is not. In the presence of calcium ions, found in abundance in rivers and the sea, the phosphate ion is precipitated out, leaving only very small amounts in solution. The usual problems apply regarding the dehydration (condensation) process which joins the components to form a nucleotide (see glossary). In living matter enzymes catalyze the process; how easily might catalyzation occur outside cells?

Chemical manipulations have produced a few nucleotides. This is negative enough, but polynucleotide strands are required which have to wrap around each other to form the regular DNA helix. This means a 100 per cent exact fit. Dr David Watts writes:

The difficulties in the way of prebiotic polynucleotide synthesis, however, are far, far greater than those which relate to the origin of polypeptides. Our own laboratory experience in the synthesis of phosphorous polyelectrolytes, which are comparatively simple analogues of the DNA main chain, leads to a vivid awareness of the need for rigorous control of monomer purity and of reaction conditions. Only with a well-designed apparatus and intelligent experimental planning can one achieve successful productions of these high molecular-weight polyelectrolytes.[7]

Perhaps this all helps to explain why DNA and other substances like it are virtually unknown outside living cells. They are products of cells and instrumental in making new ones; they do not occur anywhere else. Can anyone seriously suggest that they might have arisen piecemeal from primordial waves, when what we know of them insists that the living cells came first? DNA is exclusively a cell product. We cannot manufacture it; the closest we come to it in test tubes is to synthesize simple, short chains of mononucleotide RNA; that leaves us a long way to go. But the smallest cells of the simplest living organisms trade in it all the time. It masterminds the thousands of chemical reactions that, strictly coordinated in space and time, work together toward self-maintenance and reproduction of the cell and the whole body.

The functions of the cell, like those of a motor car, are physical, and it should be possible for us, ultimately, to understand them fully in terms of the laws of physics and chemistry. But cells and organisms are also informed life-support systems. The basic component of any informed system is its plan. Here, argues the creationist, *an impenetrable circle excludes the evolutionist.* Any attempt to form a model or theory of the evolution of the genetic code is futile because that code is without function unless, and until, it is translated, i.e., unless it leads to the synthesis of proteins. But the machinery by which the cell translates the code consists of about seventy components *which are themselves the product of the code.*

'The ribosomal system,' write Dixon and his colleagues, 'under which we include DNA and the necessary co-factors, provides a mechanism . . . for its own reproduction but not for its initial formation.'[8]

To repeat, codes and codification represent the opposite of chance. Randomness in any code sequence destroys the code. In fact, code sequences and randomness, like fire and water, are incompatible. They do not mix. And if there do exist chemically preferred and therefore non-random sequences, these are absent or overridden in biological DNA. It is therefore hugely paradoxical that some scientists suggest randomness could have given spontaneous birth to code sequences as super-specific as those of the genetic code.

Several other queries arise from this train of thought. There is no way that bases, sugars and phosphate groups naturally float together, link up, then go forth and multiply. The creationist believes that, as aluminium is right for aircraft, so it will be found that the qualities of DNA are most suitable for the design of high-fidelity organic data banks. For him, the biological silicon chip represents a molecular archetype.

After all, why should such unlikely building-blocks arrange themselves in order to produce something completely new, bearing no chemical relationship to themselves? If it was not 'in order', then protein would be a remarkable but coincidental spin-off − a spin-off whose stray strand or two would have been destroyed long before a body could form.

However 'primitive', the code must always have needed translation machinery. How, in the prebiotic ocean (or anywhere else), could the necessary battery of complex, ingenious intermediates have bobbed up simultaneously with the code to translate it on the spot? Would Darwin himself, knowing what we do, have backed his theory as far as this? Indeed, Darwin wrote that, in the beginning, the Creator breathed life into one or a few forms, which thence evolved. This is the thin end of a thick wedge because, if a single cell, the complexity of whose molecular architecture we are beginning to appreciate, could have life 'breathed into it', why not a range of more complicated archetypes?

Heavenly Genes

Laboratory experiments in synthetic chemical evolution have resulted in the appearance of amino acids, bases and sugars. This demonstrates that, under certain experimentally controlled conditions, simple inorganic compounds will form simple organic compounds. It does not demonstrate chemical evolution. Such products are as far from a living cell as the coincidence of pure latex, copper and iron would be from a functional motor car.

Sir Fred Hoyle and Professor Wickramasinghe have written: 'Biochemical systems are exceedingly complex, so much so that the chance of their being formed through random shufflings of simple organic molecules is exceedingly minute, to a point where it is insensibly different from zero.'[9]

The authors try to quantify the problem. There are, perhaps, 10^{80} atoms in the universe and 10^{17} seconds have elapsed since the alleged 'big bang'. More than 2000 independent enzymes are necessary for life. The overall probability of building any one of these polypeptides can hardly be greater than one in 10^{20}. The chance of getting them all by a random trial is one in 10^{40000}, an outrageously small probablility that could not be faced even if the whole universe consisted of organic soup.

'If one is not prejudiced either by social beliefs or by a scientific training that life originated on Earth, this simple calculation wipes the idea entirely out of court.'[10]

Certainly, as his 1981 address to the Royal Society shows, Sir Andrew Huxley was unhappy with the figures. But his appeal to a 'far simpler self-replicating system' is diversionary. There is no evidence whatsoever for any such system, the 'drive' for whose improvement into anything resembling a life-form seems, as we shall see, more in the mind of theorists than in the known properties of matter.

Francis Crick echoes Hoyle and Wickramasinghe's conclusion: 'The origin of life appears at the moment to be almost a miracle, so many are the conditions which would have had to have been satisfactory to get it going.'[11]

Following an erudite analysis of the question he suggests Directed Panspermia. If, he argues, some higher intelligence had wished to infect earth with life, it could have directed a missile loaded with compact, robust, chemically versatile micro-organisms. His choice would have been blue-green algae (prokaryotes) and yeasts (eukaryotes). Yeast is better known for the evolution of carbon dioxide than humans and, as Crick admits, regarding the origin of life we often find 'too much speculation running after too few facts'.

Spectroscopic analysis reveals that the basic chemicals of life have long existed on an astronomical scale in interstellar dust clouds. Hoyle believes that large stores of genetic material also exist, deep-frozen, in space.

'Genes are to be regarded as cosmic. They arrive at the Earth as DNA or RNA, either as fully-fledged cells, or as viruses, viroids or simply as separated fragments of genetic material. The genes are ready to function when they arrive.'[12]

How did they arrive on a likely planet? Once created, the genes travelled like seeds in the wind, riding the galaxy on the pressure of light waves from stars at speeds of up to several hundred kilometres per second. Some were caught up by comets wheeling deep in space and, in this way, were transferred to solar systems such as our own. Consequently, the terrestrial problem is not the origin of the genes but their assembly into permissible, functioning biosystems. Assembly could occur at any suitable site in the cosmos.

Both Hoyle and Wickramasinghe disavow all shackles of theology; nevertheless, they believe that Undirected Panspermia is not a matter of chance. Behind the cosmic dissemination of genes there is a purpose – to populate an otherwise desolate space. The weasel word – purpose – has appeared.

To involve purpose is in the eyes of biologists the ultimate scientific sin. . . . The revulsion which biologists feel to the thought that purpose might have a place in the structure of biology is therefore revulsion to the concept that biology might have a connection to an intelligence higher than our own.[13]

Genes are programmes. Because the evolution of life at each suitable site in the universe cannot be supervised, it is suggested that the Intelligent Programmer would 'build cells with simple robust programs, like yeast cells, that can manage to survive and multiply in the broadest possible range of environmental conditions . . . with the potential to switch over (once they have become established) to a variety of more complex programs'.[14]

This requires far more storage and subroutines on the cell DNA than initially needed, enabling it to assimilate and operate complex new blocks of information. Successful assimilation of such material would show up as a macro-evolutionary 'jump'. Genetic fragments rain gently from the heavens, programmed to metamorphose a fraction of cells by directing them: 'Stop what you are doing now, add these further instructions to your programme and then continue at such and such a point.'[15]

The carriage of such genes to their effective site in sex cells poses problems no biologist will be slow to seize. Evidence is adduced to show that bacterial spores exist in the interstellar dust clouds and may be the agents of gene dispersal. This idea is quite distinct from the origin of the first bacterium, which is not understood. Adoption of grounded DNA fragments would mostly result in pathogenic disorder or fatality. Occasionally, however, an evolutionary 'saltation' might occur.

The *facts*, Hoyle claims, have driven him to call the universe 'intelligent'.[16] His theory improves the overwhelming odds previously quoted against purposeless 'creation' of life by chance. Against the Darwinian 'theory of no intelligence', direct Creation annihilates the odds.

Notes

1   Cloud, P., 'The Origin of Life', *Nature*, vol. 296, 18 March 1982, p. 198; see also Cloud, P., 'Palaeoecological Significance of the Banded Iron Formation', *Economic Geology.*, vol. 68, 1973, pp. 1135–43; and, for contrast, Dimroth, E., and Kimberley, M., *Canadian Journal of Earth Science*, vol. 13, 1976, pp. 1176–80.
2   Hull, D.E., *Nature*, vol. 186, 1960, pp. 693–4.
3   Crick, F., *Life Itself*, MacDonald, 1982, p. 51.
4   Eigen, M., and Schuster, P., *The Hypercycle*, Springer Verlag, 1979.
5   Dixon, M., *What is Life?*, Intervarsity Fellowship, 1959, p. 6.
6   Dixon, Tipton, Thorne and Webb, *Enzymes*, Longman Green and Co., 3rd edn, 1979, p. 656.
7   Watts, D., 'Chemistry and the Origin of Life', *Life on Earth* (Journal of The Biblical Creation Society), vol. 4, p. 21, 1980.
8   Dixon, Tipton, Thorne and Webb, op. cit., p. 659.
9   Hoyle, F., and Wickramasinghe, C., *Evolution from Space*, Dent, 1981, p. 3.
10 ibid., p. 24.
11 Crick, F., op. cit., p. 88.
12 Hoyle, F., and Wickramasinghe C., op. cit., p. 31.
13 ibid., p. 32.
14 ibid., p. 108.
15 ibid., p. 109.
16 Hoyle, F., *The Intelligent Universe*, Michael Joseph, 1983.

10 Life's Engine

Metabolism is body chemistry. It encompasses the methods by which bio-chemical products are built up (anabolized) or broken down (catabolized). In the mild chemical condition of a body, such reactions are catalyzed, step by step, by suites of enzymes. Enzymes are proteins coded for by DNA: therefore, metabolism is genetically controlled.

Energy metabolism, like reproduction and sensitivity, is a basic characteristic of every living organism. Just as a motor engine converts fuel into energy, which drives the pistons, so everybody's life is linked to enzymic networks of power. These convert food into energy which, like the explosive power of petrol in a motor car, is released in a controlled and efficient way to do the body's metabolic work. In fact, a body's fuel consumption is more than twice as economical as a motor car's.

The energy a cell needs to survive has been called its cash-in-hand. Energy, which is used in all sorts of ways, must be 'ready'; it must be transferable and transformable; its use must be compatible with life-systems; and there must exist a form in which it is stored or banked. A steady income is essential to maintain, in a very controlled way, the (unnatural) form of chemical disequilibrium which is life. Bankruptcy means equilibrium means death. Because it fulfils all the above criteria, *chemical* energy is the most suitable form of energy for living cells to use. Let us take a closer look at the oxidative breakdown (catabolism) and build-up (anabolism) of food materials, which supply energy to the body, and try to account for the presence of the complex metabolic pathways they exhibit.

ATP − adenosine triphosphate − is a 'high-energy' compound which, in living organisms, forms the universal currency of energy exchange. ATP is the link between energy supplier, for example, food in the form of glucose, and the energy consumer − the cell. It is a 'charged' form of storage battery. With three phosphate radicals, high internal-energy levels hold ATP together. When enzyme action snips one of the phosphate groups off, energy is released and the molecule 'sinks' to a discharged

151

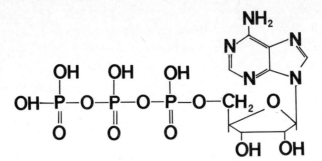

Figure 18. *ATP – biological cash* This molecule is the universal currency of energy exchange in living organisms

condition, called ADP (adenosine *di*phosphate).

The whole of life's fuel business revolves around the regeneration of ADP to ATP, recharging the minuscule batteries. As well as adenosine triphosphate, the universal biological fuel, there exist several closely related triphosphate systems which similarly store and liberate energy. Such power systems, with their accompanying networks of enzymes, must have existed in any first 'protocell'.

DNA is a chemical, but is it one that, by itself, inevitably gives rise to life? Similarly, must ATP have emerged from a pre-biological sea and, round about itself, evolved energy metabolism? No oil well ever generated a motor engine; moreover, you may remember that the component parts of ATP (adenine, ribose sugar and phosphate) are not likely to have occurred in such a sea. In fact, ATP is only synthesized in biological power units which involve the complex interaction of DNA-coded enzymes. Must a cell's capacity to use ATP have evolved in parallel with the capacity to make it? In a prebiotic sea, why should natural selection have preferred and accumulated an unstable molecule like ATP? However if, like high octane petrol, it was designed as fuel to drive a machine, its presence makes biological sense. It would mean it was created, like any part of a machine, deliberately.

Reproduction (chapter 7) universally employs DNA with its associated genetic metabolism, and one of three types of genetic division. Does energy metabolism exhibit standard components? Do compact, efficient biochemical units underwrite the power of life? We find that they do. Just as the reproductive systems of plant and animal kingdoms reflect each other, and those of male and female complement each other, so photosynthesis and respiration interlink to drive life's cycle forward. Furthermore, just as prokaryotic fission is elaborated in eukaryotic mitosis

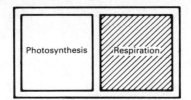

Figure 19. *Life's engine*, whose complementary archetypes are further examined in fig. 28

and meiosis, so more powerful eukaryotic systems parallel bacterial photosynthesis and respiration.

In order to solve his energy crisis, man seeks to harness the unlimited cascade of light. All the while, silently, two biological modules catch, convert and exploit the radiance of the sun. Together they make up the *engine of life*. Although its mechanism may be complex and subtle, an engine must be robust in construction. It must be able to take some knocks, although there are well-defined limits beyond which its essential working structure is destroyed. So it is with the wholesome pair, photosynthesis and respiration, on which the biology of earth immediately depends. A few mutations could occur but, argues the creationist, you would expect the fundamental units of energy metabolism to persist, working in the way they were first intended.

The Harvest of the Sun

We save or spend. Our bank account rises or falls. In agricultural terms, we save when we harvest crops: the crops themselves harvest the sun. Nature's prime source of energy is sunlight, which plants convert to food 'accounts'. These, in turn, can be cashed as ATP.

At the hub of power is a process called photosynthesis. In this, large molecules such as glucose are built up from water, carbon dioxide and sunlight – the transparent trinity that ultimately provides us with our sustenance. The opposite process – spending as opposed to saving – is respiration. Cell respiration is the way a cell derives energy.

Anaerobic respiration (glycolysis) produces a little energy, carbon dioxide and, according to the organism, wastes like lactic acid or alcohol. No oxygen is required.

Aerobic respiration needs oxygen. Occurring in special organelles called mitochondria (fig. 24), it generates much more energy, carbon dioxide and water. The intricate process is linked to glycolysis by a large, complex enzyme and a molecule called acetyl coenzyme A. This molecule 'injects'

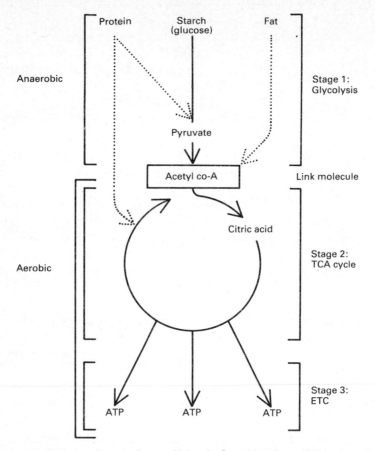

Figure 20. *Respiration:* a schematic diagram. Using the financial analogy, ATP has been likened to the body's 'ready money'. It is regularly cashed from the current account — sugars. Fats can be likened to a deposit account, to be drawn on when the need arises. Only in the case of bankruptcy are proteins, like personal property, sold to meet the body's need for energy

carbon fragments from what was originally glucose or a fat into a cycle called the TCA cycle. At each turn of the wheel more fragments are 'ground down' into carbon dioxide and a little more energy and materials are projected into stage three. Here they 'hit the jackpot', cashing ATP like coins from a fruit machine. A regular winner.

Stages one, two and three are intimately, purposively linked. In stage three electrons are tossed, like water down a mill-race, down a line of cytochrome molecules. The energy so created is used to 'recharge' ATP from ADP; by the end of the process one glucose molecule may have yielded as much as thirty-eight units of cash-energy.

If this seems a little technical and hard to understand, imagine the process as an energy 'cascade'. Energy is released when the large molecules are 'consumed' as they tumble down the cascade. They are oxidized or 'burnt'. As they are broken back to carbon dioxide and water, the energy they release is taken up by ADP to produce ATP. ATP is the 'small change' which finances each 'uphill' chemical reaction in the body. The power-mill which produces it (steps two and three especially) is fuelled in photosynthesis by the sun. In this way, that prodigal star underwrites our every energy debt.

The cycle can be pictured in terms of a reservoir and a dam. The water can, as it falls, be carefully regulated so that it performs the maximum amount of useful work. This will occur if a graded series of dams and sluice-gates afford a step-wise release of energy. Then the power can be employed to turn waterwheels or electrical generators or whatever else the designers of the hydroelectric scheme require. To 'set the wheels in motion' requires only a starting operation – opening the top sluice-gate. The dammed-up water is glucose, a molecule rich in potential energy. A little ATP is used to 'open' the metabolic pathway down which this molecule runs as it is systematically broken up. As in all systems, some energy is lost as heat due to 'friction of the working parts'. But the

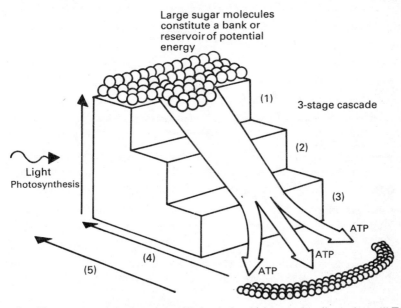

Figure 21. *The energy cascade*, in three stages: (1) glycolysis, which occurs in cell cytoplasm; (2) TCA or Krebs cycle, which occurs in mitochondria; (3) ETC. (4) is water, (5) carbon dioxide; these small molecules represent 'used energy' which is recycled to fuel photosynthesis

effectiveness of this catabolism is really cost-accounted in terms of the net production of ATP molecules.

What happens at the bottom to the unwanted end-products of the cascade? They are thrown out − carbon dioxide as we exhale, and water (if the body has no further use for it) principally as urine. But this is only half the story. Photosynthesis is like an electrical pump: it takes the still, spent water at the bottom of the cascade, and the escaping carbon dioxide, and links them. It pumps them back uphill so that the reservoir, after its depletion, is 'recharged'.

The Electrical Pump

With reason life's energy metabolism may be viewed as an electrical circuit, with photosynthesis as the battery. This battery is constantly recharged by light. Life is 'wired' lightly to the sun. To put it another way, light is like an umbilical cord by which life is attached to a cosmic dynamo − the sun. Photosynthesis takes place in chloroplasts − small bodies in the cytoplasm of plant cells. Let fig. 22 enlighten you.

Each lens-shaped chloroplast, no larger than one hundredth of a milli-metre across, has an ordered structure which seems exactly right for its function. Its chlorophyll harvests the radiant energy of sunlight and stores

Figure 22. **The chloroplast − a fuel factory**

(a) *A chloroplast* First recorded by the seventeenth-century plant anatomist, Nehemiah Grew, they have since been found to consist of a double membrane enveloping a watery matrix. In this matrix, called the stroma, are embedded a number of grana, which are made up of groups of disc-shaped vesicles stacked on top of each other like coins. These 'coins' are called thylakoids and are in turn made of thin layers, the lamellae, built of proteins and lipids. On this fine base are set light-catching gems − the emerald chlorophyll and other pigments. As precise as a solar panel, the function of a granum is to present these pigments in the optimum position for absorbing sunlight, laying them out as though on shelves, stacked with the greatest economy of space.

(b) *The porphyrin 'ring of power'* − a schematic representation. This is basic in energy metabolism.

(c) *A chlorophyll molecule* − a schematic representation.

(d) *The porphyrin head of a chlorophyll molecule.* Chlorophyll ($C_{55}H_{70}N_4O_6Mg$) is an exact arrangement of 136 atoms used to harvest light. Chlorophyll (used by plants) and bacteriochlorophyll (used by bacteria) both incorporate a porphyrin ring, although their chemical structure differs in two or three points of detail. Bacterial photosynthesis is an anaerobic process in which no oxygen is evolved

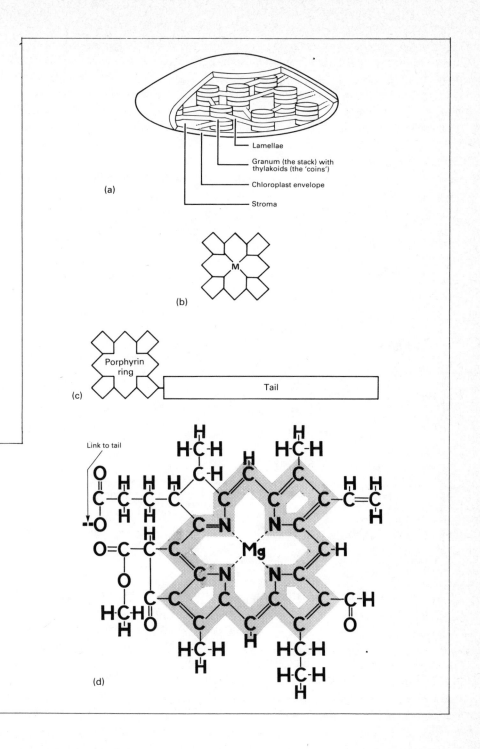

(a)

Lamellae

Granum (the stack) with thylakoids (the 'coins')

Chloroplast envelope

Stroma

(b)

(c)

Porphyrin ring

Tail

(d)

Link to tail

it as chemical energy in the bonds of sugar molecules made by the plant. Chlorophyll is made of a ring structure, called a porphyrin, and an alcohol. The ring, at whose centre a metal ion can be set, is central to energy metabolism: the structure of chlorophyll, haemoglobin (p. 166) and the cytochromes (p. 162) are based on it. Yet, like the nucleotides of which ATP and DNA are made, porphyrins are unlikely 'sailors' in a prebiotic sea: their origin is unknown.

Has chlorophyll, with its ring of power, ever existed outside a cell? The same arguments apply here as to all products of DNA coding. Chlorophyll, whether it occurs in plants or bacteria, is a complicated molecule of no use on its own. It is a product of coding synthesized along the same pathway as the red-blood pigment, haemoglobin, up to a certain point; then it receives a magnesium atom and phytol whereas haemoglobin requires instead an iron atom and protein. These 'brother' and 'sister' molecules play an integrated role in life on earth. Chlorophyll 'sparks' oxygen release; haemoglobin sweeps it to the cell's combustion chamber.

Chlorophyll works in an intriguing way. Absorption of a suitable amount of red or blue light causes the pigment to vibrate and eject an electron, which can follow two pathways. One – phosphorylation – produces a single ATP molecule. In the second system (which bacteria do not possess) water is split using the combined energy of several photons; oxygen is evolved and two ATP molecules are generated. This reaction occurs only in light. Its products fuel a second chain of reactions; these can occur in light or dark conditions and involve the synthesis of sugars, amino acids, fats and glycerols – the plant and the animal world's supply of food.

As food factories, chloroplasts are important organelles. If plants are likened to bank accounts, it is the effortless industry of chloroplasts that makes money without trying. Where did they come from? Labyrinthine phylogenies have been evolved, at least in the minds of men, to account for the origin of chloroplasts and their opposite number in energy metabolism, mitochondria. Present-day photosynthetic bacteria cannot produce oxygen and there is no proof that their ancestors could have been responsible for the oxygenated atmosphere which, with a single known exception, is poisonous to them. Blue-green algae produce oxygen but chloroplasts do not occur in them; instead, thylakoids lie free in the cytoplasm, many per cell, varying in arrangement and shape. Darwinists might say that a mutation must somehow have encased the thylakoids in serried ranks to form the membrane of the chloroplast in higher plants.

In fact, 'simple' blue-green algae are complicated. They photosynthesize and respire almost like higher plants. The fact that some can, if need be, photosynthesize using hydrogen sulphide or respire using carbon dioxide makes them *more*, not less sophisticated. In addition, some can fix nitrogen from the air, so that their food requirements are minimal. They in turn serve as a primary food source, plankton, for animals. But what was their origin? If they were anaerobic they would have been poisoned by their own photosynthetic waste, oxygen. If they used only hydrogen sulphide, they could not have served as the evolutionary source of an oxygenated atmosphere. If they were aerobic, the prebiotic 'reducing' atmosphere would have killed them. Sediments build up around modern blue-greens and cause stromatolites (p. 139) just like those from the Warrawoona cherts in Western Australia, estimated to be 3.5 billion years old. Did the most complex prokaryotes come first? Or did ancient conditions resemble modern?

It needs to be remembered that even the 'simplest' form of photosynthetic apparatus can scarcely be mimicked by contemporary scientists. Nature's copyright is simply her complexity. And, even in a prokaryote, a metabolic unit which can store sunlight in the form of starch is useless without the complementary (and equally complex) unit to use it. Photosynthesis *and* respiration, the whole energy cycle and not just part of it, is required and is present. We have to explain how this could have happened by chance, against the thermodynamic and kinetic odds described in chapter 16; and how this codeless metabolism came to precisely translate itself into DNA, so that the DNA could in future reproduce the system – otherwise it would have been lost in a single generation.

Plants are often, as regards their sexual nature, hermaphrodites; and both photosynthetic and respiratory power-packs service their energy metabolism. Why is it not so with us? Why don't animals have chlorophyll? Darwinians would say it was because we evolved from simple forms that long ago lost their chloroplasts. A pity, perhaps. Compared with the constant search for food by most animals, photosynthesis means easy living. It is a light-fantastic knack. If our skin was green and capable of photosynthesis, sunbathing (preferably nude) would replace eating. Not much would get done. No mother of invention impel. We would vegetate.

Energy Stores

A chain of sugar molecules is called a polysaccharide. There are three main kinds of polysaccharide investment. Starch and glycogen ('animal starch') are storage molecules, like current bank accounts. Cellulose, a capital investment, is a material of great structural importance in plants. In the plant cell wall it can act like steel rods in reinforced concrete: size for size, it is as strong. And gelling polysaccharides can give seaweeds flexibility or

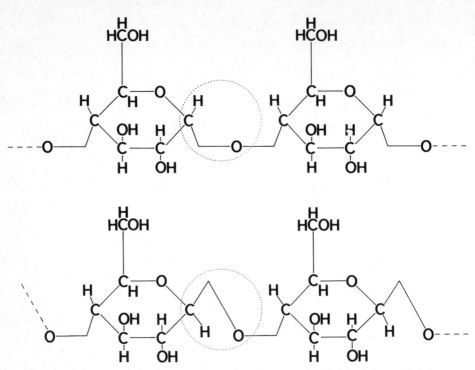

Figure 23. **Carbohydrate chains** (a) Energy for action – *starch with alpha-linkage*. (b) A shape-maker – *cellulose with beta-linkage*

act as a lubricant for cartilage joints. Proteins and lipids assume structural roles in animals; the only animals to use polysaccharides as structural components are insects and other arthropods with their tough, brittle exoskeleton of chitin.

From the creationist point of view it is interesting to note the economical 'twist' in design which differentiates cellulose from starch. The 'twist' is in the linkage by which the single sugar molecules are joined to make a chain. In starch, $\propto$-linkage (see fig. 23) results in a spiral chain which is folded and packed for storage until it is required to generate 'ready cash'. Cellulose by contrast has a β-linkage which gives rise to bundles of cross-linked parallel chains. Called microfibrils, these bundles are laid down in layers running roughly parallel to each other but at an angle (often of 90°) to those in other layers, forming tough fabrics. But even the strongest cellulose remains permeable to water and mineral solutes, which makes it an ideal structural component whose fibres are found in the cell walls of plants.

Cellulose is completely resistant to the enzymes that convert starch to glucose. Some animals, like cattle, have cellulose-digesting bacteria growing in a special stomach called a rumen so that they can live on grass, leaves etc. How, by the way, did they survive through the ages when such a digestive system was evolving?

Starch, by contrast, is a much softer material which rapidly degrades in digestion. Its linkage breaks to release glucose for energy. It is a bank associated with energy, motion, work, the body's industry and commerce. These two together, the tough fibre and the instant fuel, reflect the interesting paradox of unity in opposites which nature often shows in build-up and breakdown, matter and energy, particle and wave, fixity and flexibility, structure and dynamic process, conservation and change.

Glycolysis

Yeast cells obtain energy without the benefit of oxygen. They break down sugars by alcoholic fermentation through a complex process involving eleven specific enzymatic stages. Not until the tenth stage is any net gain of ATP realized. How could ten simultaneous, beneficial, additive mutations produce ten (or even five) complex enzymes to work, in sequence, on ten highly specific substances? Just one of these enzymes (called glyceraldehyde phosphate dehydrogenase) consists of four identical chains, each having 330 amino acid residues. Could that lot have arisen by natural selection − along with the other nine enzymes?

How can we account for complex regulatory mechanisms which control a pathway that every yeast cell can manage? There is an enzyme which limits the rate of glycolysis, and another which converts glycogen (a sugar polymer found in animals) to single glucose molecules so that the process may begin. (In more complex organisms there are hormones such as adrenalin, insulin, glucagon, etc., which control the utilization of glucose. How did these control mechanisms evolve?) In addition, a cofactor called NAD is essential in glycolysis and fermentation. Without it the processes would grind to a halt. A different enzyme is required for each process so that NAD can be properly recycled. In the absence of continuously recycled NAD, anaerobic ATP 'cash flow' would be impossible. No business is that simple.

Even suppose that Hull was wrong (p. 140) and glucose concentrations built up in the primeval 'soup'. Would such complex ways of breaking down the glucose have arisen spontaneously? Would not evolution, with its apparently inexorable drive towards efficiency and perfection, have found simpler ways? It can be argued that the present system is probably the most efficient possible. But what selective advantage could its components have had before they came together and were incorporated in the system?

Anaerobic respiration (stage one in the energy cascade) releases some energy from sugars but leaves a lot more tied up in its end-products – the most popular of which is alcohol. If you want more energy you have to breathe oxygen. In aerobic respiration glucose is converted to pyruvate which, instead of alcohol, forms an interesting link molecule called acetyl coA. Fats, too, can be converted to acetyl coA, after first being broken down to fatty acids. What happens then is shown in fig. 20. This is a much more complicated programme of events than we saw for anaerobic respiration, but the result is easy to understand. Much more ATP is produced, which is why so many organisms rely on aerobic methods for obtaining their 'cash' energy.

Several possible oxidation pathways exist but all of them present the same levels of problem to evolutionists. Even the primordial cell must have been a miniature 'superlab' in which the most sophisticated, critical and purposeful series of interrelated reactions was carried out. Where did the coded information and precise retrieval systems arise to guide and control these labyrinthine pathways for releasing energy? How could this information possibly have been marshalled *before* the cell chemistry as a whole became functional?

The Cash Dispenser

Unlike a fruit machine, each time a glucose molecule is inserted into its metabolic slot the 'energy cascade' can shute, from phosphorylation on the inner mitochondrial wall (fig. 24), its 'prize' of ATP. One of the molecules involved in this shute is called cytochrome c. Either in respiration or photosynthesis, electrons must flow and special proteins are needed to accept them and pass them along the chain. Cytochrome c is found in prokaryotes and eukaryotes wherever electrons are transferred in this way.

Cytochromes contain a haem group almost completely enfolded in a protein chain (fig. 27). Chains from different classes of bacteria and eukaryote are folded in a similar way and make a strikingly coherent family of related proteins. Within the folds, the haem group is a porphyrin ring surrounding a central iron atom, to or from which electrons can move freely. This iron atom, poised like a spider at the centre of its web, is the focus of the electron-carrying system. The haem is cradled rigidly within its protein framework by four chemical bonds. In all kinds of organism it always assumes an unexpected, energy-expensive position which R. E. Dickerson describes as a nice example of specialized molecular engineering – an

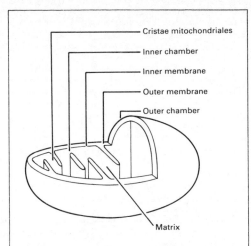

Cristae mitochondriales

Inner chamber

Inner membrane

Outer membrane

Outer chamber

Matrix

Figure 24. *The mighty mitochondrion* – a power station. Mitochondria are the sites of aerobic respiration. They mint biological cash, called ATP, from the breakdown of sugars, fats and even proteins. In its matrix, stage 2 of the energy cascade occurs; on the inner walls, called cristae, are situated the 'cash dispensers' of stage 3. Similarities between repsiratory mitochondria and photosynethetic chloroplasts are evoked in fig.28

inherently 'less than optimum' structure built into the molecule to carry out a specific function.[1] This function, electron transport, is itself part of a larger system whose rationale is to give a body energy, to support its life.

Powerful techniques have been developed by men like double Nobel-prizewinner Fred Sanger, working in Cambridge, which permit the elucidation of the exact structure (in 3-D stereo) and amino acid sequences which make up various proteins. Those for known proteins can be filled in an 'atlas' the size of a thick telephone directory. Proteins which have been sequenced for a number of different organisms include insulin, haemoglobin, lysozyme and cytochrome c.

We know that a set sequence of amino acids causes a protein to fold up automatically into a unique shape. This shape is the key to its biological activity. Many segments of the cytochrome chain, which enable it to 'pocket' haem in the right way, are invariant. Other sites, occupied by only one or other of a pair of very similar amino acids, are practically invariant: but variation between species can occur at sites which are not critical to the folding pattern.

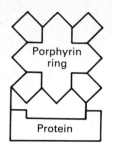

Figure 25. *Cytochrome c* – a schematic representation

Whereas in the past scientists used visible homologies, such as the penta-dactyl limb, to help classify organisms, now protein sequences from different species are compared and differences used to compute evolutionary divergence from a common ancestral gene. Take, for example, positions 62 to 84 in the cytochrome c of five organisms – man, fish, fungus, flowering plant and photosynthetic bacterium. Each letter stands for an amino acid.

Man and chimpanzee	DTLMEYLENPKKYIPGTKM--IFVG
Tuna fish	DTLMEYLENPKKYIPGTKM--IFAG
Baker's yeast	NNMSEYLTN PXKYIPGTKM--AFGG
Cauliflower	KTLYDYLENPXKYIPGTKM--VFPG
Rhodospirillum fulvum	AMLTKYLANPKETIPGNKMGAAFGG

What are we to make of this data? The segment from positions 70 to 80, the active heart of the molecule, is practically invariant. Other segments are dissimilar: in their case to interpret the difference between two DNA sequences as a measure of the distance in time by which the two organisms are removed from a common ancestor, is to demonstrate evolution only if you first assume it. The creationist also expects organisms close in the biological hierarchy, such as man and ape, to be chemically as well as anatomically close: and by degrees removed from a fish, yeast, cauliflower or bacterium.

At critical points the architecture of a molecule cannot be altered without destroying its function, its purpose. In a histone (H_4), which helps to pack DNA into a chromosome, the protein differs at only two points between mould and man! Differences in less critical parts could result from created adaptations or later mutations. The argument from molecular homology is the old argument from classification (p. 40) writ small. It is weak, like all arguments from homology, because it assumes what it seeks to demonstrate – evolution. In fact, it demonstrates creation no less than evolution.

The *origin* of cytochrome c remains mysterious. Did haem and protein

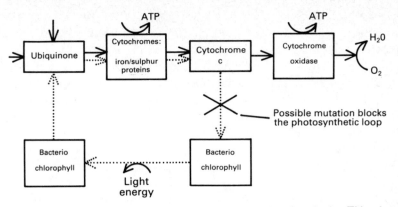

Figure 26. Some purple bacteria can carry out both *photosynthesis and respiration*. This schematic diagram shows how a single electron chain can pass electrons to bacteriochlorophyll and cytochrome oxidase. If the photosynthetic loop was blocked it would leave a chain like that found in respiring mitochondria. Arrows indicate the direction in which electrons flow

appear, with critical parts intact, at the same time and by accident? If they did not, the non-functional 'proto-parts' must have succumbed to natural selection. If they did, they would be useless alone and uncoded in a primordial soup. Not bits and pieces but the whole respiratory or photosynthetic system must arise as the consequence of specific coding. No system, no life.

Appeals to unknown simpler systems are vain. There is a limit to how simple a machine can be: this critical minimum must be present right from the start. The origins of photosynthesis and respiration are unclear. Both contain cytochromes but does this indicate common ancestry any more than the presence of magnets in both generators and electric motors indicates that one evolved from the other? Nevertheless, it is argued that respiration might have evolved from photosynthesis. Bacterial photosynthesis is anaerobic but one sort of purple bacterium (*Rhodospirillum*) can grow aerobically in the dark. As it respires its photosynthetic pigments fade. Its electron chain (fig. 26) can switch from photosynthetic to respira-

tory production of ATP and back again. Where a mutation had added cytochrome oxidase, if another blocked the photosynthetic cycle aerobic
respiration could have evolved. As oxygen levels in the early atmosphere
rose this creature would not have been poisoned. Are we, therefore,
descendents of a sort of purple bacterium?

It is not so simple. We have to account for other components as well as
electron chains which are geared into the ATP production-line. Even the
'simple' yet unique energy metabolism found in a salt-addicted bacterium
called *Halobacterium halobium* needs a large enzyme called ATPase.
ATPase has to be set in the right place, wedged across the cell membrane, before it will dispense any 'cash'. Enzymes, DNA, proteins. . . In
Israel's wilderness did evolution compose, at a stroke, the parts of a
bacterium? Was it by the Dead Sea that, with a brilliant flash, it struck on
life?

For the creationist, life's cash is not dispensed by accident. The molecular architecture of cytochrome c is functional: it is an archetypal component of life's engine. Its critical parts, which constitute strong evidence
for design, were from the first encoded. Natural selection would, of course,
preserve them intact: it would preserve the purpose of the molecule. No
life lasts five minutes without ATP and its associated power industry – so
how (on earth) could the power of life have evolved gradually?

Haemoglobin

The arguments that applied to the treatment of cytochrome c as a product
of evolution apply to all intercalated, codified macro-molecules in their
metabolic systems. Take haemoglobin. Right at the bottom of the 'energy
cascade', after cytochrome c, it is necessary for an oxygen atom to take two
hydrogen atoms and form water. This water constitutes a waste product of
the respiratory 'cascade'. If the two hydrogen atoms were not collected, the
chain would be obstructed and energy release inhibited. With no energy,
death would ensue. How is the oxygen atom brought to the site of its
important errand?

A drop of blood contains about a hundred million red cells. Each of these
small doughnut-shaped discs is covered with one of the largest and most
complex molecules in nature, haemoglobin. Its atoms ($C_{3032} H_{4812} N_{780} Fe_4$
$O_{87} S_{12}$) are arranged in four haem groups, each wrapped in a protein chain,
called globin. These chains twist themselves into an exact sculpture which
is, superficially, meaningless but in fact enfolds a trap for oxygen. A single,

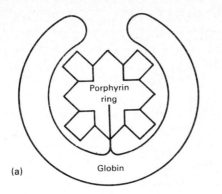

Figure 27. **The molecular lung – haemoglobin**
(a) A schematic diagram of a single-chain globin, emphasizing the oxygen-binding site. This is called a haem group and is constructed principally of a porphyrin ring.
(b) A three-dimensional representation of the single-chain globin, showing how the protein chain enfolds the haem group.
(c) A three-dimensional representation of the four-chain globin, haemoglobin

critical mutation can distort the sculpture and cause severe anaemia (p. 55). Haem, which we met in cytochrome c, is a web of about a hundred atoms arranged about a central iron atom. It is the iron to which oxygen is attracted and temporarily attached, 'rusting' the blood red.

Myoglobin is a single-chain globin, well-fitted for its job of storing oxygen for use in muscles. Could such a molecule have evolved into haemoglobin? Did the latter, with its 574 amino acids, evolve by mistake when, by gene duplication, the simple chain was doubled; when the twin pair diverged to the present situation in which they are more different than alike, and even encoded on different chromosomes; and when, finally, each twin was again duplicated by a genetic mistake, making four in all? For the single chains of myoglobin to associate, four to the larger haemoglobin complex, its primary structure would have had to undergo alterations at many points. Molecules, like machines, show critical minima stripped below which they will not work. Twenty-six critical amino acid positions dictate the shape of haemoglobin, and all twenty-six would have had to appear at once for it to function and acquire selective value. Substitution at any one of these sites destroys the functional shape. Such changes simply could not happen at once: however haemoglobin originated, it could not have evolved slowly and 'deliberately', by chance, from myoglobin-like ancestry, for the intermediate stages would have formed loose aggregates of single chains without the advantageous, cooperative interactions of haemoglobin as we know it. Neither 'good' myoglobin nor 'good' haemoglobin, how would their possessor have fared in the selection stakes?

As Max Perutz at Cambridge discovered, haemoglobin is a molecular lung. When one of four globin chains receives an oxygen molecule, its structure is altered in such a way that the remaining three chains take up oxygen more readily. This facilitates efficient transport of oxygen, especially since the reverse process occurs when one of the four units gives up an atom. So haemoglobin itself 'breathes'. Yet, like a single wrong component in a motor car, a single amino acid in the wrong place can destroy the efficacy of this molecular engineering.

As with cytochrome c, globin sequences have been worked out for many vertebrate species, and for a few invertebrates and plants. From these data there has also been an attempt to derive phylogenies, and the same criticisms as before can be applied. Haemoglobins occur sporadically among the invertebrate phyla, in some worms, some starfish, molluscs, crustaceans, insects and even bacteria. Many crustaceans use haemocyanin for oxygen transport, but because the water-flea, *Daphnia*, uses haemoglobin we do not reclassify it out of the crustaceans. No haemoglobin (but a form of 'anti-freeze') is present in the ice-fish which functions well in waters at approximately 2°C. In the sea-mouse, haemoglobin is found protecting the main nerve cord, in that its affinity for oxygen ensures that the cord will not lack that element.

Are people related to peas? Haemoglobin appears in the root nodules of peas; a nitrogen-fixing bacterium, *Rhizobium*, allows the legumes to express haemoglobin genes which normally remain unexpressed. Such genes have been isolated and shown to be very similar (homologous) to human haemoglobin. *Rhizobium*, too, has been found to contain genes for haemoglobin but of a different sort from that of the legumes. What on earth is haemoglobin doing in either peas or bacteria? Does its *Rhizobium*-triggered appearance in the legumes indicate that much genetic material, previously presumed redundant 'nonsense', can yield sophisticated products? In a similar way, stinging nettles produce histamine and other substances also found in man. Except where he is stung, are man and plant closely related?

Red Blood and Green Leaves

Power-supply units are central to all organisms. From the start of life, intricate enzyme systems have rehearsed their routines, practised their faultless choreography. DNA acts as the database from which this labyrinthine dance of life is directed.

The dance is as cyclic as the seasons. Build-up in spring is followed by the breakdown of stored carbohydrate at harvest-time. In plants both metabolisms, photosynthesis and respiration, occur. For you and me, eating is a kind of harvest-time: for the fuel of our respiration we depend, ultimately, on plants which, in their turn, depend on the sun. The chain of life leads back to the sun.

A close and complementary relationship can be drawn between the two sides of energy metabolism. The Yin-Yang concept of economically designed, interlocking opposites works well. There is, however, no hint of the origin of chlorophyll, the cytochromes or haemoglobin; nor the metabolic units in which, at molecular level, they work. Respiration (as fig. 28 shows) can be explained as reverse photosynthesis and the mechanisms of chloroplast and mitochondrion, as integrated as male and female in the sexual process, as two key factors in the whole power-supply system.

This system relies on more than precise molecular architecture, enzyme pathways or organelles. For example, modifications to the photosynthetic programme allow plants to best 'make hay while the sun shines' in differing temperate, tropical and desert climates. Organs such as stomata (plant 'mouths'), leaves and veins may be required. The same goes for the respiratory side. Molecular archetypes such as haemoglobin, cytochromes and organelles such as mitochondria are, in isolation, meaningless. They form *part* of a whole organism which includes an oxygen-delivery system, such as lungs, heart and a circulatory system, and digestive and excretory tracts. Only in the context of the *whole* organism does energy metabolism, this time at phenotypic level, make sense. Only in the same way do its origins make sense.

Euglena, a single-celled eukaryote, can use its chloroplast to photosynthesize. In the dark it loses its chlorophyll but, if supplied with nutrients, can live and reproduce. *Ochromonos*, an organism similar to

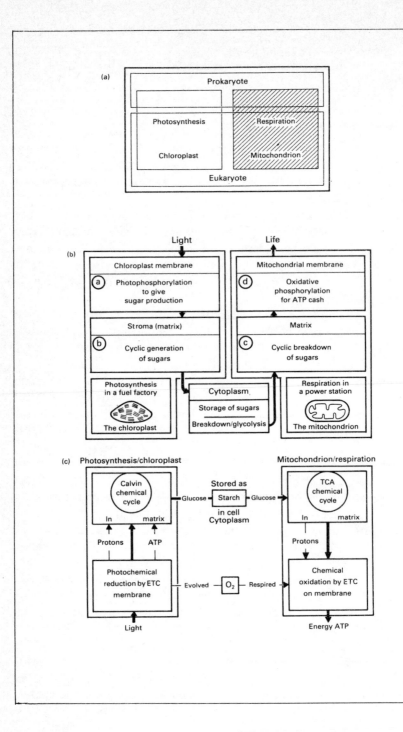

Figure 28. **The light to life energy conversion chart**
(a) *Complementary energy units* incorporated into life's engine
(b) *Chloroplasts and mitochondria* are, like a dynamo and an electric motor, the complementary opposites which interlock to drive life's engine.

The point of a chloroplast, which is invested with its own DNA for the synthesis of crucial proteins, is to reduce carbon dioxide to glucose, i.e. to make life's fuel. (a) The light phase of photosynthesis fuels the sugar-producing dark phase. It occurs across the complex inner membrane of the thylakoids (fig. 22a). Water is split into an oxygen atom, two protons and two electrons. Light falling on chlorophyll excites the electrons for use on an ETC, which makes ATP. Meanwhile the two protons are pumped inside the thylakoid membranes where they are prepared, with ATP, to drive the dark phase. The oxygen from water is evolved as a waste product. Chlorophyll, with its porphyrin ring, is the initial reaction-centre in energy metabolism.

In the dark phase of photosynthesis, sugars are produced. (b) It occurs in the colourless, fluid stroma. Carbon dioxide is taken up as a raw material. ATP and protons help drive the Calvin cycle from which sugars, acids, fats etc. are synthesized.

The point of a mitochondrion, which is also invested with some genetic autonomy, is to oxidize glucose and other food materials, i.e. to burn life's fuel in an efficient way.

Oxidative phosphorylation finally generates body-energy, as ATP. (c) The first step is when protons are pumped, by an ETC, to the exterior of a convoluted inner membrane (fig. 24). As they return ATP is made. Two 'spent' electrons, two protons and an oxygen atom then combine to make water. Haemoglobin, with its porphyrin ring, carries oxygen to the site of electron pick-up. Each atom of oxygen collects the protons and electrons to make water. The other waste product of respiration is carbon dioxide, which haemoglobin carries away. In this way, where chlorophyll initiated, haemoglobin finalizes energy metabolism.

The second stage of respiration (fig. 20) occurs in the semi-rigid matrix of a mitochondrion (d). Carbon dioxide is expelled as a waste. The TCA cycle serves to degrade sugars, amino acids, fats, etc. ATP is generated and protons are prepared to drive the 'cash release' system called oxidative phosphorylation.
(c) Aerobic energy metabolism: a more technical representation of the archetype

Euglena, can change from a flagellate with a chloroplast to an amoeboid form without. No animal produces food, no plant ingests solid food. So what are these 'planimals'?

In fact, there exist many sets of coloured and colourless organisms, whose structure is identical except for the presence or absence of a chloroplast. For example, *Euglena* and *Peranema, Chlamydomonas* and *Polytoma*. In terms of cell structure the non-photosynthetic 'animals' can be assigned to a division of algae, but zoologists include them among protozoa. The creationist suggests, not that they are the clue to the divergence of plant and animal kingdoms but simply that, in their case, they incorporate one or both programmes for energy metabolism in their DNA.

Chloroplasts for photosynthesis; mitochondria for respiration. In eukaryotic energy metabolism the two are coupled, intimately and basically intertwined. Theirs is rich, meaningful chemistry, exquisite down to the atomic and sub-atomic level. Sunlight is fixed; cosmic energy is caught, held and then released for work. Random input; organized output. From light and water comes food and cellular energy. From pure light, life.

The idea of polarity is well satisfied. So is the depth of complementary symmetry, in which chance can play no more part than in the rich and detailed patterns of an Islamic carpet. What's the problem, rationally speaking, in believing the two organelles, like the metabolism they support, encapsulate wholesome, unified design?

Notes

1   Dickerson, R.E., 'Cytochrome c and the Evolution of Energy Metabolism', *Scientific American*, March 1980, p. 103.

11 All Systems Go

We have examined subcellular archetypes as they appear in the patterns of molecules, enzyme 'laboratories' and organelles. What applies here, to reproduction and energy metabolism, applies to all archetypal subroutines that link into the programme of a whole organism. The smallest organism is a cell.

The chemical evolution of a proto-cell, its genetic code and energy metabolism; the evolution of a eukaryotic cell; the evolution of multicellular invertebrates, plants, and vertebrates – the closer we look the more clearly we see the difficulties these bland phrases bring in their wake.

The next three chapters inquire how the major vessels of life – unicellular types, plants and animals – might have been launched. What about the maiden voyages, when prokaryotic dinghies and eukaryotic yachts, motorboats and liners were floated 'all systems go'?

Monera (prokaryotic bacteria and blue-greens) are simpler in structure than protista (unicellular eukaryotic organisms), and non-photosynthetic bacteria are simpler than the photosynthesizers. What is simpler than monera?

Simplest of All

In 1898 Martinus Willers Beijerinck, a Dutch botanist, recognized that something smaller than bacteria could cause diseases. A virus is little more than nucleic acid – DNA – clothed in a protein mantle, with no cell membrane. It used to be thought viruses were very simple because they can be obtained and isolated in crystalline state. But, microscopic as they are, viruses are composed of hundreds of thousands of molecules, including over ten thousand, in exact order, comprising the DNA; the remainder are protein. The nucleic acid may become separated from the protein sheath and go into 'hibernation'; when it does it no longer acts as a virus and will not 'come to life' unless the correct protein cover returns.

Are viruses 'alive'? They possess the properties of species and they reproduce. But they lack genetic machinery of their own and have to parasitize cells of other organisms, effectively hijacking their metabolism. Different viruses are specialized to take over and exploit specific types of host-cell, usurping the host-cell enzymes. Almost all forms of life are susceptible to viral infection; we meet them, for example, when we catch cold. Some, called bacteriophages (phages for short), attack bacteria, adhering to their surface by the tail and injecting DNA in an operation similar to that of a hypodermic syringe. Having taken over the host's metabolism, the viral DNA manufactures and assembles new viruses; in as little as half an hour the bacterium may burst open and several hundred viruses tumble out.

In attack the virus is called virulent. Alternatively, as a 'provirus', it may join its host's DNA and replicate only when the host does, in which case it is called 'temperate'. But, either way, every virus needs a host-cell. There-fore its simplicity is secondary and not primary. The tobacco mosaic virus contains 21,000 peptide chains and is a hundred times simpler than a bacterium; but, like every other virus, it needs other cells to live off, and it could not have come into being before other cells were plentiful. Whence came the viruses? Were they loose genes which separated from their nuclei and developed uncontrolled multiplication? Are they mutant particles, accursed devolutions of previously healthy cells? Are they 'rogue' discords – cacophonous insertions into the symphony of life? We do not know, but they do not represent the origins of life.

The Life That Follows Death

The gap between a virus and a fully operational bacterium, with its life-support systems and intact but permeable coat, is vast. In between are the rickettsias, organisms which cause such diseases as typhus and dengue – more complex than viruses because they can carry out metabolic processes such as the oxidation of pyruvate (fig. 20). For this reason, some people have suggested that rickettsias are free-living 'ancestral' mitochondria. But they resemble viruses because they cannot survive outside host-cells: they are not candidates for the honour of grandparent. Nor, for a similar reason, are the small free-living micro-organisms, the mycoplasmas. These are bacteria without cell walls; because they are animal parasites they can exist without such protection and are considered devolved bacterial forms.

What kind of bacterium could have come first? Some bacteria live on hydrogen sulphide, sulphur etc., but require oxygen to respire. Even if their ancestors did not need oxygen it is not believed that much came from such bacteria except similar bacteria. We know how involved energy metabolism is. It blurs the issue to propose, in an evolutionary manner, that such metabolism 'occurred', that grandparent was a photosynthetic and/or fermenting bacterium and, as time went on, the suite of genes required for aerobic respiration 'fell into place'.

In fact, the 'simplest' form of independent life needs its capsule, cell wall and membrane, made by enzymes which can only do their job when contained by such a wall – and when accurately coded for and correctly synthesized. The bacterial nucleoid contains about 3000 genes (3 million base pairs in order). It can double itself in forty minutes so that DNA synthesis is done at the rate of more than 1000 base pairs per second. This extremely rapid process, performed by one enzyme molecule, shows a high degree of organization and control. How can the presence of this and other rapid syntheses be explained? Or the necessary presence of elaborate ribosomes and associated materials, such as t-RNA molecules, to carry out protein synthesis? These life-support systems must be present in even the most 'primitive' bacterium.

In spite of their simplicity, bacteria are remarkably fit creatures. If their divisions continued uninterrupted, the mass of the progeny of one bacterium could weigh as much as 2000 tons after twenty-four hours. Only lack of food and the accumulation of waste products, acting as poisons, limit their multiplication. The air around us may have more than 100 bacteria per cubic foot. Air currents carry them 90,000 feet into the atmosphere and their spores are thought to inhabit interstellar space. From extremes of cold, they have been found living in submarine volcanic chimneys at well over 100°C. They can eat sulphur, hydrogen and carbon dioxide. Some bacteria cause disease and death in man and animals. So do those other devotees of death, fungi. But many more are life-giving. The activities of innumerable saprophytic bacteria are of great importance to life on this planet. They rot the corpses, returning chemicals locked in dead bodies to the cycle of life, recycling vital carbon, nitrogen and other elements so that life may continue. Indeed, as we saw (p. 79) with plankton they form the basis of our biosphere.

Hybrid Ascension

Over the last century more than a couple of million generations of prokaryotes have been cultured and, under all sorts of conditions, observed. Although many mutations and new strains have appeared, no tendency to evolve into a eukaryote has been observed either in bacteria or blue-green algae. Indeed, from the point of view of cellular construction, a cucumber and a man have more in common with each other than with a bacterium. How is a prokaryote supposed to have become a eukaryote?

Inside the protistan *Paramecium bursaria* live numerous green algae. If a *Paramecium*, lacking algae, is introduced to them it ingests them, retaining some of them unharmed. The *Paramecium* either digests or ejects excess algae. As its photosynthetic partners, the algae produce sugar and oxygen which it uses; for their own part, they live protected inside the *Paramecium* and probably draw nutrients from its substance. It has been suggested, by analogy, that eukaryotes are, in reality, complex prokaryotic cooperatives of this kind. A theory of 'endosymbiotic association' put forward by Lynn Margulis suggests how bacteria might have taken in other organisms, transforming them into the organelles we find in eukaryotic cells.

Mitochondria could have originated as independent aerobic bacteria and chloroplasts as blue-green algae content to live inside bacteria. Ciliate structures might have originated from corkscrew-like spirochete bacteria that threw in their lot with other cells and used their cilia (whip-like structures of incredible firmness) to move or to waft materials by means of their rapid and rhythmical beatings. Might centrioles – part of the mitotic machinery – have arisen in the same way, for cilia, flagella and mitotic centrioles have a lot in common?

The argument is further strengthened by biochemical similarities between prokaryotes and eukaryotic organelles. In prokaryote, for example, SOD (an important enzyme called SuperOxide Dismutase) is built using iron and magnesium; in eukaryote cytoplasm, copper and zinc are employed. It is explained, not necessarily correctly, that the difference is due to later evolution of eukaryote cytoplasm.

In eukaryotes ribosomes are larger than in prokaryotes and organelles (but this could be a physically related effect – larger body, larger size). Like bacteria, mitochondria and chloroplasts contain DNA circlets with the accessories for transcription and translation. DNA sequences in chloroplasts resemble those in blue-green algae (both do the same photo-synthetic job). Nucleotide sequences similar to those in mitochondria have been announced for the aerobic bacterium, *Paracoccus denitrificans*. Indeed, an amoeba called *Peloxyma* can ferment organic substances by itself but lacks mitochondria. Instead, it can form a symbiotic relationship with certain aerobic bacteria!

But no present-day prokaryote forms even temporary symbiosis with other prokaryotes. The endosymbiotic theory does not explain the origins of nucleus, nucleolus or the special eukaryotic facilities for protein synthesis. These include the nuclear membrane with its pores and, extended from these, a labyrinth in whose walls DNA-translating machines, the ribosomes, are fitted. Nor does this hypothesis explain the different type of chromosomes found in eukaryotes, with their linear shape, split genes, complex coiling, mitosis and meiosis.

All organisms need stable sites for the enzymes involved in complex, sequential reactions such as occur in energy metabolism. In prokaryotes, the sites analogous to those for eukaryotic respiration, photosynthesis and replication occur as complex infoldings on the cell membrane, called mesosomes. This does not mean such sites evolved into organelles.

What of the supposed 'ingested' organisms? More than one kind of algae or 'proto-mitochondrial' bacterium might reasonably be expected to have been taken in, but chloroplasts and mitochondria are curiously uniform. They contain ribosomes of bacterial size and extranuclear DNA, but the DNA occurs in several loops rather than the single bacterial ring; and it is dependent, integrated with and controlled by the DNA of the cell nucleus. Nuclear DNA codes for key, irreplaceable proteins in energy metabolism.

An evolutionist supposes that the early atmosphere lacked oxygen and, indeed, that oxygen then as now was toxic to anaerobes. Why should complex aerobic metabolism (fig. 20) have evolved gradually, ready to switch on when the oxygen content of the air rose to a sufficient level? Why should the proto-mitochondrial bacterium, when it took up with its host, have lost its ability to ferment? Surely, double fermentation is better than single? Mitochondria can code for about thirty proteins. They do not, for example, manufacture their own lipids or cytochrome c. How, when we include other symbionts, could the control-genes of three or more organisms 'fuse' to form a cohesive, coordinated programme?

The creationist suggests that such organelles are sub-stations for the most crucial, oft-repeated routines of life – energy metabolism. A bacterium or a blue-green alga contain respiratory and/or photosynthetic power units; these archetypal modules are simply modified for insertion into the eukaryotic cell system.

Their semi-autonomous reproduction affords useful flexibility. Sometimes the cell may need more, at others less, of them. Some protein components can be ferried from the nucleus, others profitably constructed on site. Both chloroplasts and mitochondria develop from proto-organelles, which can be conveyed to a fresh body via sperm or egg. The fact that plant proplastids can develop into chromoplasts (which carry pigments), starch-sacs *or* chloroplasts supports the creationist idea that nuclear delegation of important activities to sub-stations is part of deliberate cell design.

The Margulis hypothesis, offering a hybrid ascension from prokaryote to eukaryote, is an ingenious extension of the 'bits and pieces' or 'collection of parts' approach. Others, such as D. Reanny, postulate the reverse – that the similarities between bacteria and mitochondria can be explained as well in terms of the evolutionary origin of free-living organisms from mitochondria.[1] Both ideas provide food for thought, but both will be as hard to prove as disprove. Considering the stack of improbabilities, each statistically a non-starter, which the evolutionist has to overcome in imagination, the creationist believes that his modular approach is much the more reasonable and attractive explanation for the origin of a cell.

All Together Now

This is the thin end of a thick wedge: if cells were constructed on a modular basis, why not multicellular creatures from distinct types of cell?

Metazoa, or many-celled organisms, are all eukaryotic. It is thought they arose from protozoa in one of two ways. After cell division, some protozoans remain together instead of separating into single cells. It is assumed that, out of a colony of undifferentiated cells, there may have evolved an organism of differentiated cells, a true metazoan. Colonies certainly exist in which differentiated cells assume various functions and have even lost their ability to reproduce singly. *Volvox* is made up of thousands of tiny flagellated plant cells, only a few of which, alone in the colony, assume reproductive functions. *Haplozoon*, apparently a metazoan, develops its reproductive units from single spores of dinoflagellate, i.e., protozoan structure, and is therefore classed as a colonial protozoan. By the same token could not man, with a single-celled flagellate sperm, be classed as a rather more complex kind of colonial protozoan? Where lies the distinction between such metazoans and protozoans?

Sponges live in water, with spicule or spongin skeleton forming a supportive meshwork throughout their body. In his book *Supernature*, Lyall Watson startles us: 'Some sponges grow to several feet in diameter and yet, if you cut them up and squeeze the pieces through silk cloth to separate every cell from its neighbour, this gruel soon gets together and organizes itself – and the complete sponge reappears like a phoenix to go back into business again.'[2] Is it therefore a colonial animal? In fact it is built on a different plan from metazoans, with a high degree of autonomy in its aggregated collar cells. Although sponges may reproduce asexually, they may also begin life as a zygote. Sponges are grouped separately as parazoa and regarded as an evolutionary dead end.

Was your grandparent a cell that divided and subdivided internally? Or a flatworm (you can find them in streams and gardens, hiding under stones)? Some biologists have indicated similarities between flatworms and the group of protozoans called ciliates. Many ciliates contain large numbers of nuclei within their single cell. If partitions arose between these nuclei, would the result be an elementary metazoan at the flatworm level of organization?

Ciliates and the smallest flatworms are in some ways similar, for example in the simplicity of their organization. Having no body cavity and associated organs the flatworms, like ciliates, lack permanent digestive, excretory and respiratory systems. The simplicity is not necessarily a product of relationship but of size. Small animals maintain such a high ratio of external surface to internal volume that these processes often occur through the external surface alone. In larger creatures, where the ratio is lowered, internal organs are necessary for these functions.

Unlike ciliates, a flatworm has a nerve net and reproductive organs. In some aspects these are complex enough to include, for example, a penis for hypodermic injection through the partner's body wall. They undergo embryonic development after fertilization. Ciliates stay single. They may exchange genetic material by conjugation, but after separation each divides to form two daughters: there is no separate embryogenesis.

Metazoa are classed in two groups according to whether their body develops, after gastrulation, from two (diploblastic) or three (triploblastic) layers of cells. Only animals such as hydroids, jellyfish, sea anemones, corals and comb jellies are diploblastic. Mesogloea, a layer of jellylike material between the external and internal layers of a diploblastic body, is considered an evolutionary dead end. All the rest of the metazoa are triploblasts with a layer of cells – the mesoderm – between inner and outer layers. Mesoderm is of tremendous significance and potential. In development, different regions of it grow into heart, blood and lymphatic vessels, kidney, connective tissue, muscles, cartilage, bone, the lowest layer of the skin and fluid-filled cavities.

Because the triploblastic body is bulkier, the mesoderm cells require a constant supply of food and oxygen, and the export of the waste products of energy metabolism. In this case, circulatory and excretory systems are necessary; the reproductive organs are also embedded in the mesoderm. The truth is that it is not known how protozoa evolved to metazoa: nor diploblasts to triploblasts – if they ever did. Evidence for such transitions is nonexistent. By taking two similar types, A and B, asking what changes would be needed for a transition from A to B, and thereby postulating intermediates, an evolutionist can exercise his theory. It does not mean it must have happened like that. Rather, the creationist believes the triplo-blast represents a radical departure in design and development, an arche-type which underwrites the existence of nearly all multicellular vessels, including you and me.

Protozoa include the most complex cells known; by now it will have become evident that the evolution of protozoa to metazoa is replete with speculation. The evolutionary tree, a vast structure that has grown in the minds of Darwinians over a century and more, is seedless and rootless, its own ancestry dubious, its own evolution based on the shakiest premises. No sure generalized ancestor means no sure trunk to the tree: no sure transitions mean no branches. And if we cannot discover a metazoan precursor, no branching can begin. Can the outer ramifications – the organisms we see and try to explain – have grown in thin air?

Have creationists a better explanation? According to them distinct trees appear from distinct archetypal 'seeds': there is a series of creationist family trees less exotic than the single grand tangle of evolution. Organisms are thus related in Generator rather than gene, in ancestral con-cept rather than in ancestral stock.

Notes

1 Reanny, D., *Journal of Theoretical Biology*, vol. 48, p. 243.
2 Watson, L., *Supernature*, Coronet, 1974, p. 149.

12 Green Batteries

It is time to turn from archetypes which we can only know by chemical analysis or by use of the microscope, to those which are plainly visible. Photosynthesis (chapter 10) is a kind of pump which pushes simple molecules, carbon dioxide and water, 'uphill' into an energy-rich condition called starch. Then, as they cascade 'downhill', life carefully employs that energy. You can visualize plants as secondary cells, constantly recharged by light, which drive the electrical circuit of life. They make sugar which fuels energy metabolism. In this respect, plants precede animals and we will turn to their kingdom first.

Is it not remarkable! If I gave you mineral water in the daytime, could you make a plant? Another plant can! But what is the origin of this extraordinary feat of chemistry which seems, in infinite regress, always to require a plant before the seed? What are the origins of this silent, busy world, upon whose green support life leans?

If the genealogies of animals are uncertain, more so are those of plants. We cannot learn a great deal from petrified plant anatomy which shows different species at different times but no real phylogeny at all. There are simply fascinating varieties of the plants we have today — some new species of course — plus many extinctions: but algae, mosses, pines, ferns and flowering plants are all clearly recognizable from their first appearances in the fossil record.

The palaeo-botanist (who studies fossil plants) is up against several difficulties. Cellulose is susceptible to chemical destruction; lignin or wood is tougher and the waxy cuticle of leaves may remain as a 'shell' long after the living contents have disappeared. Most resistant plant materials of all are spores and pollen, and these are therefore the commonest and most useful sources of plant information. Many plants are known only by the distinct 'fingerprint' of their pollen grains.

Leaves, wood, seeds and pollen are often separated even before the death of a plant. It is a major problem for the palaeo-botanist to assign specimens

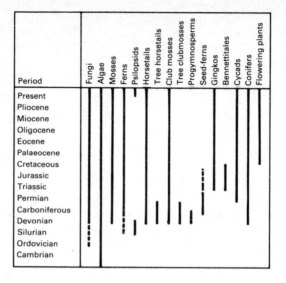

Figure 29. *A generalized geological record for plants.* Solid lines represent the duration of existence of each group. Broken lines indicate doubt as to the earliest appearance of some groups. Common ancestors are uncertain or unknown

accurately. For example, because fossil fragments tend to be given different scientific names, five names were given to specimens from coal seams which later turned out to be parts of just one kind of extinct tree, *Lepidodendron*.

Many seductive trends (chapter 7) no more indicate evolution than creation: the spectrum of land- and water-plants is one that can be interpreted in ecological as well as in evolutionary terms. Plants do, however, have one genetic trick that animals have not − polyploidy. Could this be the source of their variation?

As we have seen (fig. 8), normal cells are diploid, with two sets of homologous (similar) chromosomes. Reproductive cells (gametes) are haploid with one set. Haploid male and female reproductive cells unite in the zygote to make up the diploid number. Polyploidy goes a stage further. It is the condition, found in many plants, rarely in animals, where three or more haploid sets of chromosomes are together in the cells of an organism. Man can produce polyploid cells in plants in several ways, including the use of such chemicals as colchicine. But they also occur naturally, often persisting even if, outside cultivation, the polyploids may be less fertile and vigorous than the mother species. Some are more vigorous, making good garden and field crops. But they are incompatible with normal diploid plants within the species. This gives them a certain isolation as independent populations.

Because no mutation is involved in polyploids, neither new genetic material nor radical changes of form appear. There is, therefore, no indication that polyploidy has played a part in the macro-evolution of plants.

Soft and Cool

Whether originating from cool streams or warm seas, fossil algae are as complex as those which live today. Mosses and liverworts show their characteristic form in whatever layer their fossils are found. Because modern mosses possess chlorophyll and at an early stage in their life-cycle they look like an algal plant, it has been suggested they must have evolved from green algae. Not everyone agrees. Others argue that many algae have only one cell in their reproductive organs, while those of mosses are multicellular. And algae do not possess the sterile jacket which is necessary to keep moss gametes (eggs or sperm) from drying out in a land environment; hence their normal appearance is in water or very damp situations. The single moss embryo is housed inside the parent plant while it develops, whereas in algae the ripe zygotes are usually released from the parent into the surrounding water. The awesome, beautiful precision of these microscopic sexual encounters is worth study. We find in some respects the reproductive theme in mosses and algae is similar, in others dissimilar. No evidence exists for a leap from one to the other. A study of algae and mosses supports either theory of origins.

To suit a plant for life on land, in the dry air, more than its reproductive mechanism has to differ. In mosses and algae sap diffuses from one cell to another: mosses have small leaf-like surfaces with no protective cuticle. Because they are generally tiny, they need no internal plumbing to carry nutrients about: neither their rhizoids (roots) nor stems are vascular. Instead they have a healthy ability to dry out, bouncing back to life with the arrival of the first drops of water.

Horsetails and ferns used to be classed with mosses because they reproduce in a similar way, using alternate generations with a large frond-bearing sporophyte and small gametophyte. Nowadays (demonstrating the somewhat arbitrary and subjective nature of classification) emphasis is placed on their vascular tissue – their pipes – and horsetails and ferns are classified with the flowering plants.

It is interesting that some ferns have their xylem pipes scattered through the stem while in others the arrangement is cylindrical. The same difference occurs between the vascular bundles of flowering plants. Some, like those of the young basswood, have cylindrical bundles of tubes in the stems and scattered tubes in the roots. It cannot be argued that one structure is more 'primitive' than the other, for the stem of a plant is unlikely to be more advanced than its root.

It is believed that vascular plants (Tracheophyta) evolved separately from mosses and liverworts, in two main groups – those with the spore and those with the seed habit. From *Cooksonia*, some of whose leafless stems bore spore-cases, two major developments in land-plants are thought to have occurred.

On the one hand, a number of very different living and fossil plants are lumped together in the category *Lycopsida*. Their common element is a narrow 'primitive' leaf with only one or a few veins. These 'microphylls' connect directly to the stem of the plant. Some clubmosses bear spores in cone-like clusters: others, which are found in the Carboniferous coal measures, grew to become trees with woody trunks and branches.

On the other hand, *Rhynia*, which looks like a modern tropical herb called *Psilotum*, is supposed to have given rise to ferns, woody ferns (like a tree called *Archaeopteris*), seed-ferns, cycads, conifers and flowering plants.

In order to live on land a plant has to be able to stand upright, extract sufficient carbon dioxide from the air and mineral nutrients from water in the soil, carry this water and the products of photosynthesis wherever they are needed, and reproduce in a dry environment. For the evolutionist, the major features of land-plant architecture evolved in less than fifty million years in the Devonian period: for the creationist leaves, roots, vascular tissue, stem, spores, seeds and flowers are ready-made modules, each planned and programmed to do a botanical job of work. Each part is equally important for the whole: down in the roots of darkness hairs and ion pumps, each accumulating a certain mineral, are as essential as sunlit chloroplasts and leaves for life on earth.

Plants need to grow. A growing point, or region of active cell division, in plants is called a meristem. Principal meristems occur at the tips of roots and shoots, in young leaves and the cambium. Meristem cells develop, according to their position in the plants, into all other kinds of plant cell. Clearly, the necessary DNA for any type of cell is written into every cell. For example, each year, in a process called 'secondary thickening', the remarkable layer of cells called cambium creates new xylem tubes on the inside and phloem tubes on the outside: xylem, impregnated with lignin, becomes wood. A second layer of cambium, the cork cambium, allows the circumference of the tree's bark to increase. The origin of cambium, without which no tree can increase in girth, is unknown. For the creationist there exists a developmental programme for plants as well as animals. Meristems, including cambium in the case of trees, are archetypes incorporated in this programme. In other words, cambium is a deliberate mechanism.

Each botanical archetype is a composite of other archetypes. For example, leaf is a power station. It incorporates the soft machinery for the job of energy metabolism. Just as a dynamo incorporates such purpose-built parts as wires, magnets, coil and casing, so the leaf includes chloroplasts, guard cells, special chemicals and so on. It is well-plumbed. Even the waterproof coat, secreted in five layers with a topcoat like varnish, is complex. The location of each leaf dovetails into that of the others so that each can maximize their source of power − light. The leaf is an elegant sun-trap − although theories of its piecemeal evolution are fog-bound.

Are even the very small leaves of *Rhynia* 'primitive'? Today similar-looking leaves are quite commonly found as an adaptation to dry conditions. *Myriophyllum*, a freshwater plant, has 'rudimentary' and advanced leaves on the same stem. Submerged leaves are undifferentiated like those of various algae, while the aeriel ones are complex in structure, like those of land-plants. It is reasonable to suggest that they represent archetypes conceived for operation in water or in air.

Certainly, it is not easy to classify plants: anomalies abound. For the creationist, different plants incorporate different permutations of the archetypal modules to which we have just referred. For example, if you can imagine a fern-frond that carries seeds on its leaves, you have a picture of a seed-fern. Seed-ferns are all extinct: their origins and the reasons for this extinction are a mystery, but they attained a complexity of structure which surpassed all modern plants, except perhaps some tropical climbers. We know them only as fossils. They formed an important part of the forests which produced massive coal beds, reaching a zenith of complexity in Carboniferous times. Outwardly, seed-ferns resemble true ferns, but the suggestion that they evolved from 'typical' ferns is denied by the fact that some delicate forms are present in older rocks than the true ferns. The same general anatomy is assembled around a different reproductive theme. Similarly, the little side stems of horsetail look so much like stems of the flowering beefwood tree that it is hard to tell the difference between them. But horsetails bear no seeds or flowers, and are in an entirely different group.

Where Did the Flowers Come from?

The cycads and maidenhair tree (ginkgo) are classified as gymnosperms but, instead of wind-blown pollen grains, they have large, free-swimming spermatozoids. These, like the male gametes in mosses and ferns, swim to the archegonium and egg cell. Fossil imprints of the maidenhair's fan-shaped leaves are found as far back as the Devonian. There is no sign of evolution here, for leaves over two hundred million years old appear identical with modern ones. Most of the species of maidenhair are extinct; indeed they served as index fossils for their strata until one was found alive. Now, brought out from China where temple priests tended them in sacred groves, they adorn streets and gardens all over the world.

Because of their microphylls, clubmosses are classified separately from other vascular plants. However, one sort, *Selaginella*, has two kinds of asexual spore whose prothalli develop within the spore wall. The female prothallus has food reserves to support a growing sporophyte. Both male and female spores are found in the same cone: the larger female megaspore is thought to anticipate ovule and seed, and the male microspore, pollen. No one, however, suggests that *Selaginella*, or any fossil member of the clubmosses, was ancestral to gymnosperms or angiosperms. Waterferns and quillworts, from different classes of plant, also have such male and female spores. Oarweeds, whose female gamete is protected on the parent plant, are an algal parallel. Evolutionists say this reproductive strategy arose in otherwise unrelated plants by 'parallel evolution', but the creationist identifies it as the mosaic (non-evolutionary) distribution of an archetype.

A slow and silent explosion of greenery arises in the form of a cycad. Just as seed-bearing ferns outwardly resemble true ferns, so cycads look like ferns or palms. They occur in Jurassic rocks, for example on the Yorkshire coast: and nine genera still survive in the tropics where their unbranched stems grow to sixty feet high. Cycads are to be distinguished from tree-ferns, which have modern tropical representatives which also grow to about sixty feet.

Like pine, spruce and extinct *Lepidodendron*, cycads bear cones. Their simple unisexual structures differ from the conifers, which have both male and female members on the same tree. Examination of fossil cycad cones allegedly over a hundred million years old revealed pollen grains identical with those obtained from the male cones of the extant genus, *Anstrobus*.

Plants called *Bennettitales*, which became extinct at the end of the Cretaceous, resembled cycads but possessed superficially flower-like reproductive structures. Again, similar gross structure surrounds a different reproductive theme. In one, called *Williamsonia*, the cone was bisexual. But *Williamsonia* is considered too specialized to be a common ancestor of flowering plants. Neither were cycads nor seed-ferns ancestors because they have only unisexual reproductive structures, a feature considered to have evolved *after* bisexual ones. Some guess that the floral ancestor was a kind of magnolia. Whatever the case, the basic reproductive parts and processes of a seed-machine are worth careful study. A flower is a composite of archetypes. Were these archetypes – and not the variations on their themes found in different kinds of flower – created or evolved to ensure the plant-egg, a seed with its embryo, a flying start in life?

The development, as opposed to the evolution, of a flower on its stem is triggered by a chill on the tip of the shoot or by changes in day-length. Such changes affect a light-sensitive pigment, phytochrome; as a result a factor called a florigen is made in mature leaves and passed up to the shoot apex, whose youth is thus transformed to the loveliness of a flower. At least, that is the theory. In fact, no florigen has been isolated so that both the present development and past evolution of flowering plants remain, as they were in Darwin's time, 'an abominable mystery'.

The Birds and the Bees

Do you call it 'co-evolution' when two organisms, from different kingdoms, cooperate? Often, in insect pollination miraculous intricacy and split-second timing play their part. For example, the Looking-Glass orchid, is visited only by male wasps of one particular species. The males emerge a month before the females, and it is during their month that the Looking-Glass orchid flowers. The lip is surrounded by a fringe of red hairs just like those on the body of the wasp, and there is a bluish mirror-like patch in the centre of the lip which bears a close resemblance to the shimmering wings of the female wasp. Consequently, the male wasps are attracted to these flowers and pollination results.

Many other startling examples of pollination are known. Suffice it to mention that not only insects but molluscs, bats and birds are vectors employed in cross-pollination. Flowers pollinated by birds are generally scentless because birds have no sense of smell. They are common outside Europe and Northern Asia. What these flowers have lost in perfume has been transferred to the colour, grace and splendour of their tiny vectors. Hummingbirds, sunbirds and honeysuckers pollinate trumpet- and tubular-shaped flowers. The necks of these flowers are so narrow that only the long, thin beaks of such birds can reach the pollen. How should such specialized dependence be of evolutionary advantage?

The idea of creation is primary because it states that functional design (in this case, flower shape, beak shape, double-troughed tongue and flight capacity) was designed. The idea of evolution is secondary because it states that such design was accidental but, to explain such a self-contradictory notion, cannot render a stage-by-stage account of any but the most superficial accidents. In short, it offers an explanation without sense or substance.

Hummingbirds get their name from the whirring sound made by their wings as they move (by what stages the product of accidents?) about seventy times a second. Without feathers, the smallest of nature's helicopters is hardly larger than a bumblebee. Native to Cuba, it grows two inches long. Emerald, violet, fiery red and glowing orange – tropical quickfire flashes from these tiny souls!

Odd Couples

The phenomenon of 'convergence' of form and function in nature is so widespread that it leads to hopeless confusion in tracing relations between living forms. The parallel development of identical structures and/or biochemical systems in widely diverse types tests the evolutionist to the uttermost, although it is readily comprehensible in terms of creative permutations upon themes.

Even phyla, primary divisions of the animal kingdom, can 'converge'. For example, Hummingbird moths (also called Hawk or Sphinx moths) are equipped to hover like hummingbirds in front of flowers and suck nectar. The moth sips through its proboscis.

We can go further. Even kingdoms 'converge'. A dramatic form of 'convergence' or 'permutation of creative theme' is mimicry. Mimicry is an acute form of superficial resemblance which occurs for a definite purpose between genetically unrelated forms of life. The colours, markings, shapes and probably odours and tastes of such 'convergent types' are for protection, pollination or, as with whale and fish, function. But it is hard to explain how random mutations engender such exact and artistic replicas as you see, for example, in stick insects and mantis. These are so formed that different parts of the insect correspond to different parts of the plant it imitates. The walking-leaf insect of Asia shows a striking resemblance to a green leaf while the Mexican-leaf butterfly mimics a leaf in rest and in the way it drifts down to alight.

Startling is the life-history of the Common Mime, a butterfly from Sri Lanka. The golden egg is laid on shoots of a plant of the same order as the laurel. The young larva feeds on the tender leaves. Until half-grown it is coloured brown and yellow, with smeary-looking marks of cream colour, and has a wet-looking gloss. It always sits on the upper surface of the leaf and looks just like a bird-dropping, from which resemblance it doubtless benefits.

Next, from half-grown to fully grown larva, when it is too big to imitate a dropping, it assumes a gaudy black, yellow and red coloration. Such gaudiness is usually associated with poisonous and repellent qualities. By advertising these qualities the insect is, to a degree, protected from attack.

When the caterpillar turns into a pupa, the insect returns to the device of resembling an inedible object and assumes the form of a short, snapped-off dead twig. The base of the pupa is so shaped that it appears to grow out of the stem to which it is fastened, and the far end is shaped to resemble the broken-off surface of a twig.

The emergent adult butterflies are of two separate forms, males and females occurring in both. The type which is brown and mottled with yellow mimics the common brown Euploea butterfly; that which is striped black and blue mimics a species of Danais butterfly. This is not all. Both forms of the Common Mime have distinct styles of flight. The one used when they are alarmed or pursued is the ordinary swift, dodging flight of the swallowtail butterfly to which the Mime belongs. A slow, sailing, careless type of flight used by both Danais and Euploea butterflies is adopted by the Mime when feeding and playing, and by the female when laying her eggs. So alike are the flights of mimic and mimicked that even a trained observer is hard put to distinguish between them. In this case the reason for mimicry is that the Euploeas and the Danais, with their warning colours, are known to be distasteful to birds. Many other startling examples of accurate mimicry exist. Where did each originate? Could co-evolution, or any other sort of evolution, have caused the Common Mime?

13 Programmes Typed in Dust

Surely fossils record metazoan evolution? They do, although the information can be negative. Palaeontology, the study of ancient forms of life, does not inevitably mean the study of evolution.

Cambrian rocks exhibit an explosion of life, so termed. All kingdoms and subkingdoms are represented in the geologic record from here onward. So are all classes except vertebrates, insects, moss corals, and the extinct trilobites and graptolites. Divisions of the plant kingdom, except algae and fungi, appear later. No new phylum (or division) has arisen for 350 million years and there is a complete absence of transitional series between any two phyla. In excess of 5000 species adorn the Cambrian layers.

Cambrian animals include sponges, corals, jellyfish, worms, molluscs, trilobites, crustaceans, in fact every one of the major invertebrate forms of life. Nor is the record entirely biased. Marvellous imprints of both soft-bodied and shelled animals occur in the Burgess shales of British Columbia. Complex organs such as intestines, stomachs, bristles and spines are found: eyes and feelers indicate the presence of nervous systems. Many life-cycles show alternation of generations, for example, the medusae and polyps of a jellyfish. Nematocysts are specialized sting-cells that develop in the bodies of such creatures. They contain coiled, thread-like harpoons. These poisonous darts are explosively triggered and entangle the hapless prey, which is then drawn by tentacles into the hydroid mouth. What is the step-by-step origin of such a unique, involved weapon-system? There is little justification for writing off the Cambrian fauna as 'primitive' if 'primitive' implies 'half-formed', 'inadequate' or 'simple'. Darwin himself admitted that the situation was inexplicable and could be urged as a valid argument against the views set forth in *The Origin of Species*.

Thick sections of sedimentary rock are known to lie in unbroken succession below strata containing the earliest Cambrian fossils. The Green shales of Britanny (17,000 feet) and the Huronian series in Canada

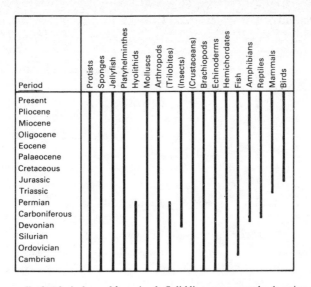

Figure 30. *A generalized geological record for animals.* Solid lines represent the duration of existence of each group. Broken lines indicate doubt as to the earliest appearance of some groups. No common ancestry is certain

(18,000 feet) are two examples. Although pre-Cambrian limestone deposi-tions are not short of calcium, a great number of species did not acquire the secret of hard-shell development until, at the base of the Cambrian, wave after wave of phyla appear with shells or skeletons. These include arthropods, brachiopods, molluscs, echinoderms, sponges and, by the end, corals and chordates.

Although lack of hard shell is one of several reasons put forward to explain the paucity of pre-Cambrian material, there is the *Cloudina* anomaly. Calcareous tubes of this species (an annelid worm?) have been lifted from the pre-Cambrian in Namibia. How could this be, millions of years before the widespread radiation of hard parts?

Nowadays, thanks to modern research, the moratorium on knowledge of the pre-Cambrian has been lifted. Nevertheless, pre-Cambrian is only a name. Fossils are used as the key for placing rocks in chronological order. The criterion for assigning fossils to specific places in that chronology is the assumed evolutionary progression of life; the assumed evolutionary progression is based on the fossil record so constructed. The main evidence for evolution is therefore the assumption of evolution. Allied with radiometric dating assumptions, which fundamentalist Christians dispute, is the fact that frequent contradictions to the evolutionary succession occur

and are 'explained away' by overthrusts, unconformities etc. The fundamentalists are busy cataloguing such *non sequiturs*, for evolutionists would
rather argue them away than take them as serious indications of error.

Leaving aside geological doubts, there sometimes occur doubtful identi-
fications. Desiccation curls, as representing scales of dried mud, may be
interpreted as arthropod shells. A discovery was claimed by a US
geological survey team in 1972 of arthropods in pre-Cambrian rocks from
the Sierra Ancha area of northern Arizona, dated at 1.2 billion years – for
the evolutionists several hundreds of millions of years too early. Pyrite
rosettes may seem like jellyfish; drag-marks made by water drawing fragments of limestone crust over a surface might be construed as worm-tracks.
Is *Brooksella* a bubble of gas in sediment, a sort of jellyfish or a trace fossil
of egested sediment after a meal outside a worm burrow? When you cannot
even tell a phylum from a trace fossil, you reach the limit of the record's
worth.

Leaving aside doubtful evidence, prokaryotes have been found from
very early ages. Fossilized algae are, in almost every case, similar to
present-day algae, indeed all major groups are found in the lowest
Cambrian rock. It is, though, one thing to discover bacterial or algal micro-
fossils and another for these to 'explode' into such diversity as the billions
of Cambrian fossils exhibit. For this reason, it is argued that metazoan
history must stretch far back into the pre-Cambrian in order to account for
the number of different body-plans, i.e., archetypes which the Cambrian
displays.

In 1947 in the Ediacara hills of south Australia a geologist, R.C. Sprigg,
came upon varieties of fossil jellyfish in sandstones designated lower
Cambrian. The examination purported to show that the fossil-bearing
rocks were pre-Cambrian. Here was a pre-Cambrian fauna rich in
complex, multicellular life. There were many soft-bodied specimens,
including jellyfish, sea-pens, segmented worms, spicules (possibly from
corals or sponges) and a couple of extinct oddities. Except for the oddities,
the animals bear a close resemblance to living types. Nevertheless, the
current interpretation of the Ediacara fauna represents a relief for evolution and was celebrated in an article by M.F. Glaessner.[1] The Ediacara bed
is an outcrop, formed from a depression supposed to have been an inland
sea. It is because the surface rocks which fill this basin have, without
radiometric calibration (see chapter 16), been designated Cambrian that
the lower system has been termed pre-Cambrian. The sceptic queries such
chronological capers. Since layers are usually identified by the fossils in
them, should the stratum not be called (as it originally was) basal
palaeozoic, i.e., Cambrian?

It remains an enigma how bacteria evolved into jellyfish, sea-pens or worms. Even evidence of shelled molluscan evolution, the best, is speculative. Only theory commands that prokaryote evolve to eukaryote, or single-celled eukaryote into a multicellular 'metazoan'.

The Fly on the Wall

'God', said biologist J.B.S. Haldane, 'must have possessed an inordinate love of beetles.' Eighty-five per cent of metazoan species are arthropods. Of these, 78 per cent are insects; there are three-quarters of a million species, of which beetles make up a large proportion. The scientist was gently mocking creationists of his day, but the plethora of beetles is a problem for evolutionists too. There are three main kinds of arthropod: crustaceans such as shrimps and water-fleas, chelicerates like spiders and scorpions; and uniramians which include *Peripatus* and insects. The characteristics of the three kinds are so different that it is impossible to derive any one from another. So it is hypothesized that the 'common ancestor' was a soft-bodied worm-like creature. This, in turn, evolved three ways into hypothetical hard-bodied crustacean and chelicerate ancestors; and a soft-bodied insect precursor. None of these hypothetical creatures is known. They are simply intelligent inventions to 'plug the gaps'.

All Cambrian arthropods, except the extinct trilobite, fall easily into modern classes. *Rhyniella*, the earliest known insect, is from the Devonian sandstones of Rhynie, Scotland. It is now accepted that *Rhyniella praecursor* belongs to a living order which includes Collembola or springtails. Indeed, such is the similarity between fossil (380 million years old) and modern forms, with their specialized locomotive apparatus from which springs their name, that 'contamination' was suspected. This suspicion has been dispelled. Ancient is ancient, but still just like modern.

Insects have always been numerous and varied. As many types live now as in the past. Did they evolve from the caterpillar-like worm, *Peripatus*? It does indeed possess features sometimes found in worms or insects; other features are found in neither. All three types of creature were present in the Cambrian period so that evolution from annelid (worm) to arthropod must have occurred, if at all, in very early Cambrian or pre-Cambrian. Of this there is no evidence. The three occurring, fully formed and complex, together in the Cambrian leads to the belief that transitional forms never existed.

Insects allegedly appeared, able to metamorphoze and moult, long before the first vertebrates. Maggots will more or less dissolve themselves when developing into a fly. Was the process pre-programmed from the first 'production run'? Or was the ancestral fly a dissolved maggot?

Insects appear to have remained in equilibrium with their environments. Amber fossils indicate, for example, that even the complex social habits of ants have not changed for 35 million years at least. Samples show plant lice attended by ants in search of honeydew and the presence of mites attached to ants in the same manner as today. In their other forms also insects have 'marked time' in an evolutionary sense. Why?

The helicopter in a thimble (p. 225) shows that insects possess some striking survival systems. Creationists often point to the bombardier beetle. Under attack it aims two tubes in its tail at the enemy. There is a miniature explosion and fumes are projected which effectively repulse such small predators as ant, spider, frog or praying mantis. The components of the mechanism include two adjacent storage sacs, combustion chambers, gun-like swivel tubes and, of course, the bombardier's instinct for using them. Chemicals (hydroquinones, hydrogen peroxide and enzymes) are present in the right places, amounts and concentrations. A series of nerve and muscle attachments co-ordinate the system and aim the protective spray. This spray is not continuous. Each discharge, pulsed as a rapidly firing machine-gun, can be heard as a 'pop'. An almost instantaneous reaction takes place in which oxygen blasts out a quinone spray under high pressure. Space rockets work on the same principle. As well as boiling, the beetle's spray is toxic and malodorous! The millipede *Apheloria corrugata* shoots hydrogen cyanide at aggressors! How does it not poison itself?

Any such complex mechanism must, as a gunner will understand, be fully coordinated and functional before it is the least use. If not, the bombardier beetle would have blown itself into extinction, or at least boiled itself alive. By what stages could it have evolved? What proof is there of such stages being more than theoretical imagination, such explanation being correct? Or does the bombardier beetle blast evolution?

A Fishy Invertebrate

There is no fishy invertebrate: transition from invertebrate to vertebrate supposedly passed through a notochord stage. The notochord is a skeletal rod, a hollow dorsal nerve cord arranged along the head-tail axis of creatures in the phylum chordata. The chief invertebrates have an exoskeleton, dorsal circulatory system, ventral nervous system and muscles inside the skeleton. To obtain a chordate, you have to turn an invertebrate inside out: then put skin and bones in the right places.

Invertebrate exoskeleton in the shell of a crab, for example, is made of a stiff, tough substance called chitin which is lighter and stronger than bone and serves as a suit of armour to protect the internal organs. In composition it is quite unlike cartilage or bone, being a nitrogenous sugar. As the creature grows, it moults its exoskeleton. Cartilage occurs in a few invertebrates, such as squids and limpets, but bone is unique to vertebrates. It can form by ossification of an embryonic rudiment of cartilage; or develop directly from tissues under the skin. Skeletons are feats of engineering. Each bone, down to the pattern in which material is laid down inside it, is finely designed to best withstand the stresses of its work: it employs the appropriate joints and articulations in order to fit, in a precise way, with its neighbours. Thus it ramifies into a purposeful structure. How could a skeleton gradually evolve? Although the earliest known fossil vertebrates were bony, the bone archetype occurs suddenly, fully formed in the record: it has not changed since it appeared.

Various solutions have been suggested to the mystery of the missing bones. Suspects include acorn worms, sea-squirts and curious 'calci-chordate' fossils, traditionally classed with sea urchins and sand dollars as echinoderms. But top of the list is a brainless 'fish' called *Amphioxus*. It has no fossil record and, it seems, is not on the direct evolutionary line of vertebrates but it gives a clue of what early chordates might have been like.

A cartilaginous notochord appears briefly in all chordates as an embryonic precursor of the vertebral column. Remnants persist between the vertebrae which replace it in the adult. *Amphioxus* is the only chordate that keeps its notochord, above which runs a hollow nerve cord, throughout its life. In *Amphioxus* it carries transverse bands of muscles which, contracting rhythmically, send a series of waves down its back: these push water backwards and thereby propel the creature forward.

The notochord differs in fine structure from that of other chordates. Changes in hydrostatic pressure within its sheath affect the stiffness of the organ and may also be related to locomotion. In other words, the notochord is a specialized hydrostatic skeleton; it is functional and teleological, fulfilling express locomotive purposes.

Nevertheless, if *Amphioxus* did not evolve from an Idea, the evolutionist is obliged to detect a chain of events leading up to, and away from, the suspect's back. The adult sea-squirt is a transluscent, muscle-ringed 'vase' which quietly sways and trembles, not unlike an anemone, at the bottom of the sea; but as a baby 'tadpole' it possessed a flicker of a notochord. *Amphioxus* always has one. An ammocoete, once classed with *Amphioxus*, is now known to be the larva of a lamprey; lampreys, classed with hagfish, look like soft-skinned, medium-sized eels but have no bones in them at all. They are classed as jawless fish. In this way, could the evolutionists have reconstructed a plausible scenario?

Because the embryonic form of a sea-squirt exhibits a sliver of cartilage in its back, it has been suggested that it could have traded metamorphosis for evolution: if a little mutant once reproduced without growing up, might not the entire vertebrate phylum have evolved from it? On the other hand, notice how the trochophore larvae of annelids and molluscs (whose adult body-plans are widely different) are almost identical. This suggests to the creationist that larvae be considered an archetypal developmental stage, between unprotected egg and adult form.

The unique notochord of *Amphioxus* extends over its mouth. The creature, unlike an ammocoete, is not only brainless but eyeless. Stranger still, its excretory organs (called nephridia) utilize cells which are comparable to those found in flatworms, annelids and molluscs. The vertebrate kidney cannot have its origin in these: yet, in the other direction, such creatures are remote from *Amphioxus*. Although they are unaccountable in evolutionary terms, a creationist perceives in *Amphioxus* evidence for the mosaic incorporation of subroutines into an organism. It represents a permutation of archetypes, almost as strange as those of a duck-billed platypus (p. 210).

In truth, F.D. Ommaney writes:

How this earliest chordate stock evolved, what stages of development it went through to eventually give rise to truly fishlike creatures, we do not know. Between the Cambrian, when it probably originated, and the Ordovician, when the first fossils of animals with really fishlike characteristics appeared, there is a gap of perhaps 100 million years which we will probably never be able to fill.[2]

Is God of this gap? Or can mutations plug it? If different programmes were typed in that special dust, DNA, you might expect as large a gap between any two programmes.

Jaws

A lamprey is a jawless wonder with virtually no fossil history; there is nothing like it. The earliest bony fossils were also jawless but this time were clearly fishy customers. One kind, the Heterostracans, do not seem to have had an internal skeleton, but they all had armoured plates, tails, fin-flaps etc.

If cartilage precedes bone, should not bony fish have evolved from sharks or other elasmobranchs – which haven't a bone in their body? If their cartilaginous skeleton is termed 'primitive' or 'degenerate', this only demeans the evolutionary use of such words. Perhaps the shark was always separately typed in life's indelible ink, the DNA.

Acanthodes sport a spine on the leading edge of each fin and, more importantly, a jaw. So do plate-skinned Placoderms, strange and varied creatures which swam in Devonian waters and, except for the Chimaera, are extinct. But extinction is never evidence for evolution. Placoderms fit no evolutionary pattern; no ancestors are known. Indeed, all major fish classes are clearly set apart, without transitional links. Moreover, there is no evidence for the evolution of a jaw in any jawed fish. It has been argued that two anterior gill pouches, with their gill bars, ceased to inspire and became a jaw. The ex-gills then sprouted teeth. This may explain the origin of jaws better than the origin of eyes, paired limbs or fins; but it does not satisfy the creationist, who sees jawlessness as a deliberate inhibition of full fish potential: the archetypal programme has been restricted. Who complains? Lampreys so successfully rasp the flesh, and hagfish eat right through their living prey, that they are now pestilential species.

The archetypal vertebrate programme is self-contained and complete. There is no fossil evidence to shed light on its origin. It can be modified to generate birds, reptiles and mammals as well as fish. And still further refinements, such as the four classes of jawed fish, can be conceived. As previously noted, the creationist believes that such refinements were conceived down to about the classification level, genus. Thereafter, variation was, and is, due to the deliberate incorporation of genetic flexibility – due to sexual meiosis.

Whatever their origins surprisingly few major taxonomic groups have no survivors at all. Why are 'transitional' species so rare and debatable? Why are the persistent types so persistent or, to reverse the question, why did their related species become extinct? The answer is, we do not know. But perhaps 'old four legs', the coelacanth which, at this moment, swims lazily off the East African coast, can give us a clue.

Fins

Fish like mudskippers climb trees. So can some frogs and crabs. Eels wriggle across land. It is supposed that an enterprising fish became the first amphibian. Today, if a population of fish were removed from water and thrown into the middle of a desert, they would die. But evolutionists would have us believe that drought forced fish up and away into a permanent divorce from water on to land.

The Australian lungfish has gills and a lung. The African and South American lungfish have two lungs; this pair inhabit stagnant waters and, with the help of a heart which partially separates oxygenated from deoxygenated blood, can use the lungs exclusively when their ponds dry up. As the pools shrink they dig down, curl up and line their holes with slime. When the mud dries, the slime forms a dry skin round them so they can rest for many months until the rains reactivate them. But their sex cells cannot mutate and they cannot reproduce while in this dormant condition. The lungfish has not 'taken up its bed and walked': rather, it has changed little since Devonian times. Anyway, the bones of their skull are so different from those of the first fossil amphibian that they are disqualified as 'missing links'.

The youngest fossil coelacanth is about sixty million years old. Since one was rediscovered off Madagascar, they are no longer claimed as 'index fossils' – fossils which tell you that all other fossils in that layer are the same ripe old age. Do their lobed fins represent, as evolutionists say, a step toward walking limbs? Although the pectoral fin can be rotated through 180°, there is no elbow joint in it. The pelvic (hip) bones are loosely attached in the muscle and do not connect to the backbone. Legs are rigid, with at least three stout long-bones and feet to stand on, joints at elbow and wrist, knee and ankle, and muscles and ligaments to work them. Why should a fish *need* to grow legs? The lesson from 'old four legs' is 'once a fish, always a fish'. It has not ventured from its archetypal theme.

You can't say that about *Eusthenopteron* because this fish is extinct. While the upper bones of its fins resemble the upper bones of the legs of land animals, it is equally true that the 'hand' part of the limbs is quite different. A fin carries a web supported by fin-rays that may or may not have a central bony axis. The bones of a hand or foot are quite different from fin-rays in origin and development. No intermediate is known. Nevertheless, the tiny bones of *Eusthenopteron*, said to resemble the humerus, radius and ulna in amphibians, have given rise to a selection of imaginative 'photopaintings' of this ancient Moses and its exodus on to dry land. J. Monod has written: 'A primitive fish "chose" to do some exploring on land, where it could however only move about by clumsy hops. The fish thereby created, as a consequence of a change in behaviour, the selective pressure which was to engender the powerful limbs of the quadruped.'[3]

It is as if, by hanging long enough to a cliff, the human being might evolve hook-like hands or wafer-thin toes the better to manage steep rock crevices. Descendants of Monod's fish have apparently taken other forms in a fantastic manner, fulfilling, extending and amplifying 'the dream of the ancestral fish'. Without intermediate forms between coelacanth or

Eusthenopteron, it is a thin story, but vertebrate textbooks give it the authority of holy writ.

The coelacanth has 'advanced' vertebrae, so has an early amphibian, about a metre long, called *Ichthyostega*. The three living orders of amphibians, however, (newts and salamanders, caecilia, frogs and toads) all have 'primitive' vertebrae. These orders, as different from each other as from *Ichthyostega*, are found complete from the start. The first fossil frog, modern in almost every respect, went a-wooing among Triassic rocks.

There is no evidence that fins developed into legs or, for that matter in whales, legs into fins. Although newt-like *Ichthyostega* has 'fin-bones' in its tail, it is very different from a coelacanth, lungfish or *Eusthenopteron*. It has a true neck, limbs, fingers, toes and a greatly modified skull. Since the earliest tetrapods are found in upper Devonian rocks, contemporary with the fish from which they are supposed to have descended, some unseen line is supposed to have evolved from lobe-fins. This is, as usual, theory-driven speculation.

Eggs before Chickens

There is no clear evidence as to which group of amphibians was directly ancestral to the first (or any other) reptiles. Vertebrate fossils are mostly bones and teeth and there are few skeletal differences today between certain groups of amphibians and reptiles. *Seymouria* has been offered as a link but it is possible, if we could see its internal organs and life-cycle, that it would be classed with amphibians. It has characteristically amphibian teeth and fossil larvae from a related genus have gills. Anyway, it occurs in the early Permian whereas preceding 'stem reptiles' such as *Hylonomus* (a cotylosaur) and 'mammal-like' Synapsida are found in Carboniferous strata.

One 'soft' part has been fossilized. Dinosaur eggs, for example, have been found in the Gobi desert. Amphibians and reptiles lay eggs, but reptiles lay them out of water. Nevertheless, the reptile infant needs a watery environment and an endowment of food; whereas amphibians hatch as larvae which fend for themselves, reptiles hatch tiny but fully developed. So the reptilian egg must include a store of yolk and protein-rich albumin, and be cased in a container or shell. Yolk, in which grows the embryo, is attached to the shell by elastic threads.

The shell demands delicate compromise. It must be strong enough to resist accidental breakage but fragile enough for the chick to chip free. It must lose the right amount of water so that the embryo neither dries out nor drowns in its own metabolic water. Its size and nutrient content must be geared to embryo size at birth. And gases must be able to diffuse through pores which are the result of deliberately randomized packing of its calcium-carbonate crystals.

Such a shell requires two special embryonic membranes – the amnion and allantois – to protect the embryo, allow it to breathe and act as a reservoir for the waste products resulting from metabolism. Waste products must take the form of insoluble uric acid (produced by birds and reptiles) not the soluble urea produced by amphibians and mammals.

Fertilization of the egg must occur within the female before the shell begins to harden, necessitating concomitant changes in the urogenital organs and habits of the adult. And the hatchling needs a chipping tool to develop at precisely the right time and place, along with the right instinct to chip out of its cradle. Most of these changes would be useless if not harmful until more or less complete. We are required to believe that these factors evolved simultaneously, by chance, with harmonious, interlocking functions; at the same time the whole scheme of amphibian metamorphosis must be scrapped or switched off and immediately replaced. Without all these arrangements (which are supposed to occur gradually over relatively long periods of time) the creature would die before hatching. But *if* evolution is true, it *must have* happened.

You would think amphibians had the best of both worlds. The pity is that, in becoming reptiles, they devolved their natural aqualung. Just think what you and I could have achieved with gills, as well as with lungs! After hatching a reptile has to control water loss. The skin has to be waterproofed so that it cannot be used, as with amphibians, for breathing – well-developed lungs and an improved circulatory system compensate. There is no evidence that the drastic distinction between reptile and amphibian evolved piecemeal, by chance. Equally, there is no proof that it occurred as the result of a fresh programme employing a pristine batch of subroutines in the DNA, but that seems at least as plausible an explanation.

If not in Eden. . . .

'. . . upon thy belly shalt thou go and dust shalt thou eat all the days of thy life'.

A satanic snake creeps quite unlike a legless lizard. The outcast from the garden slides, coiling smoothly due to special rolling joints. Its ball-and-socket vertebrae are supplemented by further articulations that clinch strength with sinuosity down the length of its back.

No snake has any trace of shoulder bones or a fore-leg. In pythons and boas horny 'spurs', joined to a small pelvis by a 'femur', assist in movement and, terminating externally at the cloaca, in copulation. The creationist interprets these 'vestiges', like the chick teeth and fibula (p. 224), as a serpentine modification to the vertebrate programme.

No snake bit the dust. Instead, several bones of its skull can move independently and the jaw-bones are joined to the skull by elastic connective tissue. So a large python can swallow a whole goat. Its gullet is distensible and its ribs unattached for the passage of the prey − but still the brain and skull must be protected from the struggle. So boas and pythons coil and kill. Other snakes are venomous; their sharp teeth, which point almost like an incantation towards the back of the throat, are supplemented by an additional row on the roof of the mouth. The victim is worked down the throat head-first by the alternate action of the upper and lower tooth rows. The snake cannot choke because its windpipe begins far forward in the mouth. Then the most powerful digestive juices in the animal kingdom dissolve bone, hair and hide.

What about the fangs? A snake's poison apparatus includes a chemical plant for making venom, a safe storage vessel and effective delivery system (the fangs), the instinct to use it properly and safeguards to ensure the snake does not poison itself. The fangs are sharp and delicate, with grooves or holes through which the toxin can be injected under pressure. Provision is made for their replacement; three or four fangs lie below each other on each side of the mouth, ready to be pushed forward.[4]

It is unlikely that the serpent's brain, protected at the base of the skull by an enlarged and toughened sphenoid bone, is as devious, devilish or seductive as tradition suggests. But a snake has a lot of sense. All the time its forked tongue flickers in and out, picking up small particles from the air or ground and transferring them to Jacobson's organ. This is a pair of pits in the roof of the mouth with a lining similar to that of the sensory part of the nose. Although the nose has a good sense of smell, Jacobson's organ sharpens it.

Pit of hell! At dead of night a snake 'sees'. Facial pits in the snout of the snake can detect very small differences in temperature − even that surrounding a trembling mouse or frog.

By day, the snake does not change focus by changing the shape of its lens like other reptiles; instead it shifts the whole lens. Again set apart from other reptiles, the cone cells of its eye − which give sharpness of vision and are associated with colour vision − are double. And no doubt coincidence evolved the only two transparent scales in its body to cover its eyes!

The bone which in lizards conducts sounds from the eardrum to the inner ear, in snakes attaches to the jaw. The temptor, which 'hears' vibrations on the ground through its jaw, was deaf to the airborne whispers of Eve.

Snakes undulate on land and in the sea. There is even a flying snake which spreads its ribs, flattens its body, glides between the branches of South-East Asian trees and then resumes both its cylindrical shape and its course of business. Today, snakes differ from other reptiles in the same way as their fossil record, from the Cretaceous, shows they always have.

Notes

1   Glaessner, M.F., 'Precambrian Animals', *Scientific American*, vol. 204, no. 3, 1961, pp. 72–8.
2   Ommaney, F.D., *The Fishes*, Life Nature Library, p. 60.
3   Monod, J., *Chance and Necessity*, Collins, 1970, p. 121.
4   McMurray, N., 'Ophidia', *Creation Science Movement Bulletin*, no. 171, p. 3.

14 The Mammals

Listen!

Of the dozen or so differences that separate mammals from reptiles and birds (fig. 32) the one the evolutionists have taken most to heart is the articulation of the lower jaw. In mammals a single bone called the dentary (because it bears the teeth) makes up either half of the lower jaw. It articulates directly with the squamosal area of the skull. Reptiles have a more complex lower jaw with no fewer than six bones in either half. One of the bones of the jaw (the articular) articulates on the quadrate bone of the skull, a bone not found in mammals.

All reptiles have a single rod-like bone in the ear (the columella) which connects the eardrum to the inner ear. There is no evidence that its 'simplicity' in any way impairs the hearing of its possessors, who can perceive pitch and volume as well as mammals; bird song would be wasted, at least on birds, if the columella lacked efficiency and left them partly deaf. Mammals possess three bones in the middle ear (stapes, malleus and incus, so called because they resemble, respectively, a stirrup, a hammer and an anvil); also a complicated inner ear, with the organs of balance and cochlea, containing the organ of Corti.

Where does this dry lesson in cranial anatomy lead? A good deal more than a few Mesozoic teeth and jaw fragments hinge upon it. Such fragments, interpreted as belonging to transitional reptile-mammalian stock, are supposed to demonstrate the origin of mammals. In this way they silently articulate the evolution and status of mankind.

Although modified subroutines govern the structure of ears and jaws in all vertebrates, this does not necessarily mean one kind evolved into another. Consider what such a transformation would require. Some early reptile would have scrapped the original hinge of its lower jaw and replaced it by a new one attached to another bone. Five bones of the lower jaw would have broken away from the biggest bone. The jaw-bone to

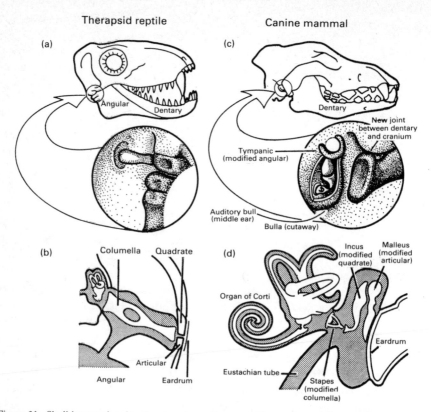

Figure 31. Skull-bones related to **the structure of mammalian and reptilian ears**. The articular and quadrate bones that formed the articulation of early reptile jaws also carried vibrations from the external eardrum through the columella to the inner ear. In mammals these functions are split: the dentary forms the whole lower jaw and efficient ear ossicles interact with the organ of Corti. (a) *Therapsid reptile.* (b) *The entire reptilian ear.* (c) *Canine mammal.* (d) *The entire mammalian ear*

which the hinge was originally attached would, after being set free, have forced its way into the middle part of the ear, dragging with it three of the lower jaw-bones which with the quadrate and columella, formed themselves into a completely new outfit.

While all this was happening two complicated structures would have developed in the inner ear. The organ of Corti, peculiar to mammals and their essential organ of hearing, comprises some 3000 arches placed side by side so as to form a tunnel. Study the complexity of the cochlea and its nervous connections. Add to this the vestibular component of balance, which includes three semi-circular canals in planes at right angles to each other. Two different kinds of nerve receptor are finely designed to achieve their purpose. Both pieces of apparatus are intimately linked by their fluid (endolymph and perilymph) systems. There is no evidence that such elaborate evolution could, or did, take place. The apparatus is entirely novel; from what precursors did it derive?

No problem? A popular school textbook gives an imaginative account of this evolution of vertebrate ear ossicles and inner ear. It ends: 'Of course, there are numerous unresolved questions about this story; for example, how did the mammal-like reptiles hear, and chew, while these fantastic changes were taking place? But despite such functional problems there is little doubt that it happened.'[1]

Little doubt! While all other changes that turned reptiles to mammals were occurring too?

Ictidosaurs, such as *Diarthrognathus*, South African fossils of mouse-like form, are offered as evidence. Palaeontology is not shy of prejudicial language and *Diarthrognathus* is classified with 'pre-mammalian' cynodont reptiles. This incomplete curiosity is said to possess both reptilian and mammalian jaw articulation because the articular surface of the lower jaw is made up partly by the articular bone (as in reptiles) and partly by the dentary (as in mammals). The upper articular surface on the skull consists of both quadrate (reptilian) and squamosal (mammalian) bones, so that four bones participate in one articular joint. It is guessed that the reptilian bones might have been involved in sound conduction in 'advanced' reptiles and 'primitive' mammals. As long as the four bones retain a function in jaw articulation, however, is it proper to call any of them ear ossicles?

The same argument applies to *Probainognathus*, a fossil reptile from Argentina whose lower jaw consisted of several bones. It is the current favourite because *Diarthrognathus*, being too specialized as regards other features, is considered an evolutionary dead end.[2]

Another candidate is a 'mouse' from Bridgend in Wales. It is called *Morganucodon*. A morganucodontid fragment was recovered with the quadrate still in contact with the articular bone. This 'mammal' had a powerful reptilian jaw. Its bones were found in scattered fragments which had to be 'reconstructed' to make sense of them: indeed, most fossil remains of mammals retrieved from the Jurassic and Cretaceous periods, allegedly spanning more than a hundred million years, could be held in your cupped hands. No intermediate fossils, with three bones in the jaw and two in the ear, have been found; nor, although the morganucodontid dentary extends back sufficiently to encourage the belief, is any fossil available showing the dentary in actual contact with the squamosal of the skull. Such contact is only a matter of inference. Even if it existed, could it be called a mammalian jaw-joint existing alongside a reptilian? Not even a single intact lower jaw is available; such specimens as exist have been reconstructed from fragments.

The creationist, of course, claims that mammalian and reptilian archetypes are 'closed circuits'. No genes ever flowed between them. So diverse is the skeletal structure of reptiles that hardly a single unique and consistent feature defines it. Some overlap in shape is possible between classes of four-legged vertebrates − reptiles, mammals, amphibians − in spite of large physiological differences. Closer resemblance than usual is not the same as evolution. It is a pity that cynodont reptiles, of whose soft parts we are ignorant, became extinct before we could precisely determine the value of their 'pre-mammalian' title.

Different but not Evolved

It is supposed that the two main divisions of mammals − the pouched marsupials and the placentals − arose from a common tiny ancestor the size of a shrew, perhaps the British *Morganucodon*, in late Triassic times. The monotremes (duck-billed platypus and spiny anteater) are supposed to have diverged before this.

By the beginning of the Tertiary period most placental groups (carnivores, insectivores, primates, rodents, rabbits, bats and hoofed mammals) were in existence. The origin of whales is unknown. As with seals, walruses, monotremes and man, their fossils occur later. The earliest marsupials appeared about the same time as most mammals and 'evolved in parallel', with carnivorous, herbivorous and insectivorous stock.

Although the first marsupial was, allegedly, of opossum-type from North America, the most diverse fossil assemblies have been obtained from South America and, later (Pliocene), Australia. It can be argued that marsupials travelled from South America via Antarctica to Australia before continental drift split these land-masses apart. They flourished on these 'giant arks' until, a few million years ago, the isthmus of Panama rose from the sea. Placentals drove across this isthmian bridge and soon displaced the 'inferior' marsupial varieties but, of course, they could no longer reach Australia. Marsupial fossils, presumably from creatures that crossed from America before the continents split apart, are also found in Western Europe.

Are marsupials worse than or, in an evolutionary sense, prior to placentals? They represent only 6 per cent of the 4200 mammal species yet they

Figure 32. **Differences between mammals and reptiles**

(i) Differences between shoulder and hip girdles.

(ii) The blood of mammals, is maintained at a constant temperature, while that of reptiles is not.

(iii) The substitution of non-nucleated red blood corpuscles for nucleated ones. It has been suggested that mammalian corpuscles have lost their nuclei to economize on material. Why has such economy not also evolved in birds and reptiles?

(iv) The main product of excretion in amphibians and mammals is urea. In reptiles and birds it is uric acid.

(v) Mammals have a single dorsal aorta (great artery), reptiles have two. Both survive successfully. How did one circulatory system change into another?

(vi) Mammals have fur even if some, like elephants, or whales, become almost devoid of it. Reptiles have scales, birds have feathers. There is no explanation for their origin and the creationist regards each covering as a separate archetype. Feathers develop from mesoderm while hair, with the exception of basal papillae, comes from the outer skin. A few mammals, like the armadillo, have hair and scales: did the 'reptilian' scales re-evolve?

(vii) The diaphragm, a fibro-muscular partition separating the chest and stomach cavities, is not found in reptiles or birds. As a result these creatures breathe quite differently from mammals.

(viii) Mammalian milk is a perfect food, on which babies can thrive for months. It has proteins for body-building, minerals for bones and teeth, fats and sugar for energy and vitamins for vitality. Satisfactory production mechanisms, storage vessels, delivery channels and the instinct (in mother and child) to suckle are present.

(ix) A secondary palate, which separates nasal from oral cavity, is present in mammals but not in reptiles or birds.

(x) Elaborate dentition, including milk teeth, occurs in mammals; reptiles have single peg-teeth and birds none at all.

(xi) Relatively speaking, mammals have larger brains than reptiles and birds.

(xii) The occipital condyle is a bony knob at the back of the skull, articulating with the first vertebra. You use it when you nod. It is absent in most fish, double in amphibia and mammals but single in reptiles and birds. It seems evolution wavered on this important point!

(xiii) Mammals have three ear-bones instead of one. The inner ear is very much more complex.

display about as wide a range of body forms and manners of living as placentals. Their pouched design is unsuitable for aquatic and aerial varieties, and these do not exist.

The marsupial brain may be slightly smaller than the placental but there is no evidence that either this or a different brain organization decrease fitness. (It lacks some connective fibres called corpus callosum.) Despite aspersions cast on their wit, marsupials have survived and flourished. The

Virginia opossum, supposed to be the dimmest of creatures, has spread through the States into Canada. It has a small brain but, in the long run, so what?

It seems that nature has been parsimonious with her basic mammalian archetypes. In some cases, such as the marsupial mole of Australia and the Golden (placental) mole of South Africa, the bodies are practically identical except for the pouch. Four unrelated groups of anteater (the ant-bear, pangolin, aardvark and spiny anteater) are found on four separate continents. Extinct marsupials exist in the form of South American borhyaenas and sabre-toothed tigers. In Australia fossils have been found of a tusked marsupial 'lion', a ten-foot kangaroo and another creature as large as a rhinoceros. Marsupial jerboas, flying phalangers, dasyures, wombats, Tasmanian devils and thylacines parallel placental jerboas, flying squirrels, cats, woodchucks, badgers and wolves found on other continents. We can compare the eating habits of the koala bear with the leaf-eating sloth or the slow loris. The large and small kangaroo species that graze the Australian plains are the equivalent of the diverse African antelopes. But there is no equivalent of the primates: no marsupial woman is known!

A powerful chapter in Arthur Koestler's book *The Ghost in the Machine* is called Evolution: Theme and Variations. It is about homology and archetypes. In a section entitled Doppelgangers he writes: 'It is almost as if two artists who had never met, never heard of each other and never had the same model, had painted a parallel series of nearly identical portraits.'[3]

The evolutionist argues that parallel forms occur either by parallel or convergent evolution. For example, the common ancestor of mammals may have been a mouse-like monotreme from which evolved placental and marsupial forms; from the latter marsupial and placental *doppelgangers* evolved in parallel. Convergent evolution is said to occur when, due to similar life-styles, differences between two or more lineages decrease. Examples are the anteaters or whales and fish.

In contrast, the creationist identifies reproductive, respiratory, nutritive and other themes in life. Allied to these he perceives biochemical, physiological and anatomical archetypes – expressed as homologies. Variations on these archetypal themes have been created to allow organisms to live in all sorts of style. The example of the pentadactyl limb for locomotion by land, water and air has already been given. For the creationist, therefore, placental and marsupial *doppelgangers* represent an archetype for body shape which incorporates two different reproductive themes. In the case of the dog archetype this results in the wolf, on the one hand, and the thylacine (marsupial wolf) on the other.

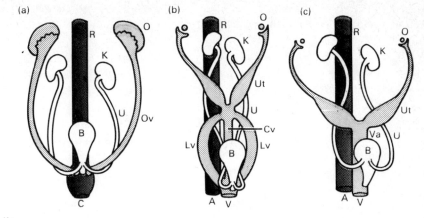

Key
R – Rectum
A – Anus
C – Cloaca
V – Vulva
Ut – Uterus
O – Ovary
Ov – Oviduct
Va – Vagina
Lv – Lateral vagina
Cv – Central vagina
Ur – Ureter
B – Bladder
K – Kidney

Figure 33. **Mammalian reproduction – an adapted archetype** The female reproductive systems of the three mammalian subclasses. In the monotremes the oviducts, bladder and rectum all open into the cloaca (a combined reproductive and excretory tract), and the large ovaries yield big, yolky eggs. Marsupials have paired vaginae and a central vagina or birth canal is formed temporarily at the time of birth. Placentals have a single vagina and, typically, a two-horned uterus. They produce tiny eggs. (a) *Monotreme system.* (b) *Marsupial system.* (c) *Placental system.*

Seen in this way, 'better' or 'worse' has no more to do with it than convergent or parallel evolution. Marsupials are not seen as living fossils in any way inferior to placentals, but as equivalent species built around a different reproductive design. The evolutionary origin of that design is, for both groups, unknown.

Anatomical status as marsupial or placental seems not to affect the survival stakes. It could be argued that there is no scrap of evidence to show that the apparent defeat of South American marsupials was due to an invasion of North American placental superiority. Some biologists prefer an ecological argument predicting hard times for any group of South American carnivores. If the types had been reversed and North American marsupials had moved South, it may still have been the South American creatures, this time placentals, which would have lost out. At any rate, this is what Stephen Gould thinks.

It has even been argued by J. A. W. Kirsch of Yale that marsupials did not just take refuge in Australia but that this was their centre of origin. Were marsupials the southern and placental the northern archetypes?

Who knows? But it looks as if the myth is exploded. One is neither better nor more advanced than the other. Just different.

What has a pouch, a sticky tongue, a pair of ankle spurs and no teeth? The spiny anteater lays eggs, has fur and behaves like a reptile. An odd mosaic indeed! Would anyone like to divine whence that 'monstrous imposture', the duck-billed platypus, evolved? It lives in the freshwaters of south-eastern Australia. With a large flat beaverlike tail, ducklike beak and five webbed toes, it swims like a fish. It uses echo location like a bat or dolphin and a hollow spur on the inside of its heel connects with a poison-gland, making it the only venomous furred creature. It is warm-blooded and gives milk like mammals. Its limbs are as short as a reptile's but it has large cheek pouches like a monkey or squirrel. Young platypuses have teeth which in adults are replaced by horny plates, unique among mammals. Such an arrangement suggests that the young monotreme, born from an egg, always needed to suckle from structures which resemble the human nipple and areola.

The bodies of an ape and a dog incorporate the same reproductive archetype in the placenta; on the other hand, in the case of a wolf and a thylacine, variations on a reproductive theme (fig. 33) can occupy similar bodies. Naturally, an egg-layer might incorporate a different system from a viviparous mother. Female monotremes have a cloaca − a common urogenital and excremental tract. So does a placental beaver. Marsupials have paired vaginae with a central birth canal formed temporarily at that time; on the other hand, placentals have a single vagina.

Whereas in the platypus (and snakes and lizards) only one ovary and oviduct − the left one − are functional, in the spiny anteater (and croco-diles, turtles, marsupials and placentals) two are functional. What pattern can we perceive? Did the spiny anteater evolve from the platypus? And then give rise to marsupials and placentals − excepting beavers?

As regards sequence data, the DNA complement of monotremes is nearer (with 97−98 per cent) that of placental mammals than marsupials (with 81−94 per cent). Monotremes have no real reptilian tissue; but from what did they evolve? What reproductive structure had the primordial mammal? Most mammals from this period are known only from jaws and teeth − the rest is reconstruction. Nor is the platypus among them; its fossil record consists of some teeth less than three million years old. It is very 'young', very 'modern'.

Until the 1960s monotremes were considered 'obviously' primitive, incompletely evolved mammals with poor thermoregulation. Then research was actually done which showed that they are as physiologically sophisticated as any other mammal. For the evolutionist they represent a baffling puzzle. Their unique combination of structural features renders it impossible that they arose from any particular class of vertebrates, could have been intermediate between any two classes or are themselves related. For the creationist they incorporate an almost climactic array of permutations, of adapted archetypes which, as two mosaics, constitute these remarkable creatures.

The ancestry of rabbits, horses, sheep, rodents, dogs, cats, cows, sea-cows, elephants, giraffes, whales, monkeys, duck-billed platypus, marsupials, sea-lions and great apes (including man) is uncertain. The creationist sees them as 'different but not evolved'.

Monsters of the Deep

It is as easy (or as hard) to show that an organism could not possibly have evolved from any other, as to show that it did. Water rat, otter, duck-billed platypus or polar bear are aquatic mammals that show no signs of evolving into seals or whales. Except for the platypus they differ so little from terrestrial mammals that, to look at them, you would hardly suspect their aquatic talents. They do not need fettered hind-legs, nor do they lose the battle for survival without them. Yet whales are believed to have evolved from primitive terrestrial mammals, their shape converging with that of a fish. Outside Disneyland there exist real problems in transformations of this magnitude. Look at fig. 34.

The only fossils we possess of early whales do not differ substantially from modern ones and we can presume that the physiological mechanisms essential to cetacean survival were also similar. Recently,[4] scattered fragments of bone (the posterior portion of a cranium, two fragments of the lower jaw and isolated teeth) were assigned to the order Cetacea. Great is the need for a transitional half-whale, but *Pakicetus inachus* − if it is all one creature − lay in an assemblage of fluvial fauna, had the auditory mechanism of a land animal and teeth reassigned from a species of hoofed land animal. Has not the mind transformed these bones? Evidence for the evolution of either toothed or baleen whales is absent.

The vertebrate structure, like the invertebrate, plant and other structures, is a vehicle for life. Like motor vehicles, a range of organisms is generated by adapting main features for use in different conditions. The whale's prototype incorporates subroutines (oral, aural, visual etc.) adjusted for use in water. Just as papyrus reeds were treated to make scrolls and circuits are printed on silicon chips prepared from sand, so DNA is a

Figure 34. **From a Mouse to a Whale?**

(i)   During a transition period a quadruped would have had a hipbone too small to support the hind legs and yet too large to permit the musculature necessary to move the great tail of Moby Dick. The whale has a small pelvis, to which various muscles are attached, and two small internal bones of the femur, which may act to strengthen the genital wall. It seems more likely that the mammalian theme is so coded in whales that it triggered these forms than that the pelvis and 'hind-legs' have, by a series of genetic mutations, for which there is no fossil evidence for any transitional stage, beneficially 'wasted away'. We need to remember that mere lack of use cannot cause hereditary change or disappearance of structure. For example, a man may never use his biceps but his genetic coding will ensure that his children are adequately equipped in that respect.

(ii) The whale's forelimbs are jointless paddles or flippers; there is no fossil evidence of evolution into these 'primitive' organs, which serve their purpose admirably.

(iii) The whale lacks hair (except on the snout) and, to enable it to keep its temperature above that of the surrounding water, the body is protected by a thick layer of fibrous, fatty blubber, a blanket needed even in tropical waters. Sweat glands are absent.

(iv) The outer skin is criss-crossed with numerous striations, like those on an engraving, which appear to help streamline the flow of water in a pattern which gives maximum speed for minimum exertion. This is a remarkable energy-conserving feature to find (randomly?) coded into the whale's DNA.

(v)   Eye, ear and mouth differ from those of land mammals. The eye, as well as several other minor but helpful changes, is subtly converted so that light rays through the sea water are brought to focus on the retina. In water man becomes long-sighted, while with the whale it is the reverse. In water it sees well but in air everything becomes blurred. It has a sclerotic coat for protection at great depths. Fine adjustment in mouth and nose design serves to prevent water entering when the giant is swimming or under pressure deep-diving.

(vi) The ear is constructed on a different plan from that of the mammal which receives air-borne sound-waves. From a narrow opening behind the eyes sounds are carried down a tube to a point where they impinge on the eardrum. This sensitive skin, set deep in the skull, is protected from the high pressure experienced in diving. The orifice is minute

material for the storage and retrieval of biological information. It is argued that Leviathan, the whale, is one of a suite of vertebrate programmes typed, once and for all, in this very special kind of dust.

for good reason but whales can communicate over long distances, so that the ear is not functionally deficient. It is able to resist temporary high pressures at depth, does not appear to have changed in degree of complexity since the earliest Zeuglodonts (ancestral whales) appeared, and is altogether remarkably attuned to the demands of its environment.

(vii) Instead of teeth the whalebone whales have great baleen plates which hang like curtains from the roof of the mouth. These act as ingenious and perfectly moving sieves or traps for plankton extraction. The blowholes are placed backwards, permitting respiration without elevation of the muzzle above water level. This is helpful for a creature with no neck (or at least an inflexible portion). Toothless whales (Mystacocetes) in the foetal stage have the germs of teeth. This feature, although referred to by transformists as 'vestigial', has a purpose in that it supports the bone as it develops into the large organ of a baleen-plated whale-jaw.

(viii) Since the whale submerges for fairly long periods (up to two hours), the breathing apparatus has been refined to ensure a sufficient supply of oxygenated blood and absorption of carbon dioxide; also to withstand the great pressures exerted by the water in deep dives. Could the whales' deep-diving equipment which so effectively resists great pressures have evolved by chance? It could not have evolved because the whale wanted or needed to dive; that is obsolete Lamarckian doctrine. Further, before it was fully functional, the creature would have been crushed in the attempt to dive. Not much food exists at great depths, so what was the characteristic selected for? Whales possess refined submarine navigational and tele-communication systems.

(ix) The female whale bears her young tail-first and suckles them under water. She secretes milk into a specially developed nipple which fits the baby whale's snout. Thus the baby is prevented from imbibing sea-water with its nourishment; and its windpipe is prolonged above the gullet to prevent milk ejected from its mother's 'breast' from entering its lungs. Choking underwater can be lethal, so this series of creative, integrated yet random mutations was vital. What happened to young whales before they had all occurred?

Notes

1   Roberts, M. B. V., *Biology, a Functional Approach*, Nelson, 2nd edn, 1976, p. 586.

2   Lillegraven, J.A., Kielan-Jaworowska, Z., Clemens, W.A., *Mesozoic Mammals, the First Two-thirds of Mammalian History*, ch. 3 'The Origin of Mammals' by Crompton, A.W. and Jenkins, F.A., Jnr, University of California Press, 1979.

3   Koestler, A., *The Ghost in the Machine*, Picador, 1975, p. 143.

4   Gingerich, P., et al. *Science*, vol. 220, 1983, pp. 403–6.

15 Ordinary Exotica

Suppose that, in the future, the combined efforts of science were able to 'sculpt' an exact, metabolizing copy of a human. Would it, moving dead, be a robot? Or would it 'come alive'?

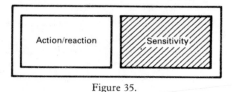

Figure 35.

You don't have to be alive to be sensitive. Machines with appropriately programmed control systems can be as sensitive as we are to things that matter for survival — temperature, chemical stimuli, sound, light etc. — without experiencing anything at all. Many complex human operations, although under nervous or hormonal control, require no conscious experience for their exercise. And, although pain serves its purpose, machines can be programmed to avoid what might damage them.

Nevertheless, sensitivity is perhaps the most profound, intangible characteristic of life. Sensitive experience is the basis, not of biology but of an organism's own life. It feels and it responds. Like a pilot it gathers information and reacts to the body's flight through time. Why, though, do we *need* to experience anything? Why do we *want* to survive? What is the origin, or the point, of conscious life?

Each religion claims to have the answer; evolution has not. This book attempts neither to analyse brain and behaviour nor to define the nature of consciousness. The senses of sight, hearing, taste, touch and scent are, like radio or TV, however, receivers of vibratory data. The translation of this data informs and integrates the individual with the world outside the mind. Of these, the most powerful register is sight. Of course, it is not known how the nervous signals which reach the brain from the eye are translated into the visual experience of surroundings. Nevertheless, we can spend a few moments examining the physical eye – just one of many complex items which developed from a code-loaded ovum not a millionth part as large.

Sight

'Was the eye,' wrote Sir Isaac Newton in his *Optiks*, 'contrived without skill in optics and the ear without knowledge of sounds?'

What if Darwin had been confronted with the anatomical and physiological knowledge of modern biology? Sir Charles Sherrington in *Man on His Nature* arises like the ghost of Paley, the creationist whose *Natural Theology*, published in 1802, made such a profound impact on Darwin. He observes: '. . . beyond the intricate mechanisms of the human eye lie breathtaking glimpses of a Master Plan.'

The eye appears in the fossil record as though sprung fully formed like Athene, the goddess of wisdom, from the brow of Zeus. On the other hand, the theory of evolution assumes that the most complicated eye, for example, that of an eagle or a man, started as a freckle or light-sensitive spot which was gradually transformed by chance in upward stages, eventually becoming a working, purposeful, complex mechanism with millions of parts. It is assumed that the light-sensitive cells were slowly folded inward to form, progressively, a retina. The skin on the surface then became transparent and turned into a lens to focus light on to the retina.

Consider the eye 'with all its inimitable contrivances', as Darwin called them, which can admit different amounts of light, focus at different distances, and correct spherical and chromatic aberration. Consider the retina, consisting of 150 million correctly made and positioned specialized cells. These are the rods and cones. Consider the nature of light-sensitive retinal. Combined with a protein (opsin) retinal becomes a chemical switch. Triggered by light, this switch can generate a nerve impulse. Retinal is an archetypal molecule, a foundation for visual sense. Each switch-containing rod and cone is correctly wired to the brain so that the electrical storm (an estimated 1000 million impulses per second) is continuously monitored and translated, by a step which is a total mystery, into a mental picture. Who, we ask, is the ghost in the machine experiencing this phenomenal image? Who are you, the seer behind the rearrangement of light?

Evolutionists employ the veil of time to blur their vision. The story starts with a light-sensitive spot, such as is found in *Euglena*; a series of broad but superficial strokes culminate in the complex eye of an octopus, or the even more complex vertebrate eye which is yours. But a list of eyes from various animals, not necessarily related, no more demonstrates evolution than a carefully ordered range of lamps. Darwin said: 'If it could be demonstrated that any complex organ existed, which could not possibly have been formed by numerous, successive, slight modifications, my theory would absolutely break down.'

The eye is a good contender. Suppose that a light-spot remained at the bottom of a cup-shaped organ and a lens formed at the top. A nerve fibre then connected the light-spot with the brain and the cup became elongated so that it could only respond to light coming from a particular direction, as if someone were looking down a tube. The compound eyes of insects consist of a very large number of these organs, called ommatidia, grouped together. The optical images are assembled from dots, each dot from an ommatidium, somewhat as a TV picture is assembled from many light and dark dots. The picture is composite.

There is the problem of how a group of ommatidia get together to make an eye. But what about colour vision? It is found in several bony fishes, reptiles, birds, bees and primates. Among mammals only primates see in colour. Dogs, cats, horses and bulls do not. Fish supposedly evolved the necessary retinal cones to give them colour vision, but then lost them. 'Re-evolved' by certain unrelated birds and reptiles, they were lost by mammals, but by luck 're-surfaced' in primates. An odd story indeed.

Given such a diverse 'mosaic' spread, it is reasonable to assume the sub-theme (colour vision) is coded, something like an 'optional extra', onto the main visual theme. Such permutations as we find could arise by this adaptation. A more likely story? Creationists think so; at least it is plausible.

An eye, like a television or camera, exists to attain an end – sight. It is teleological. All types of eye, based on the light-sensitive cell, are simply variations on the coupled theme of optical-image perception and inter-pretation. Both faculties are required; each is useless without the other. Eye, sight and meaning are inextricably entwined. It is reasonable to argue that, just as a film camera is unthinkable without purpose and intelligent information embodied in it, so an eye is the product of concept, not chance. That such an instrument should undergo a succession of blind but lucky accidents which by necessity led to perfect sight is as credible as if all the letters of *The Origin of Species*, being placed in a box, shaken and poured out, should at last come together in the order in which they occur in that diverting work.

The vertebrate eye is, in principle, quite different from the compound eye. The image is 'simple', uniform and inverted like that of a camera.

How could this type of eye have developed from the normal invertebrate type? It is no use invoking cephalopod molluscs, like the octopus and squid, whose eyes bear striking similarities to our own. They are genetically unrelated and no series, leading up to their extraordinary optical apparatus (in some ways excelling our own), exists. A squid can distinguish polarized light, which we cannot, and their retinas have a finer structure which almost certainly means they can distinguish finer detail than us.

Two sorts of eye are required. A very small eye, suitable for an insect but built in a manner similar to that of a man's eye would not work because it would not be able to diffract the light enough. Its possessor would scarcely be able to make out the shape of objects at all. Conversely, it seems that a much magnified compound eye would prove vastly inferior to an eye having lens and retina. There is teleological necessity for the two designs.

But *how* did an eye or two arise? There is no evidence for any transitional form, even if one were feasible. We are not treated to a detailed account of the evolution of retina, cornea, rods and cones, visual photochemistry, tears ducts, lids, muscles etc. Can such an irrational and hollow hypothesis be called scientific? What advantage, as far as natural selection is concerned, could accrue from the starting of an eye when the materials forming it were not yet transparent. In the human, coding generates biconvex lenses, purposely free from blood vessels, and focusing apparatus which is exquisitely refined. The eye must be perfect or near perfect. Otherwise, it is useless.

Of what survival value is a lens, forming an image, if not intimately linked to a nervous system which will translate that image into electrical form? Or a nerve without a brain to interpret the data? How could a visual nervous system have evolved before there was an eye to give it information? So questions continue until all parts of the body are woven into a single whole, a web of mutual necessity.

Darwinism does not look you squarely in the eye. It insists on faith in the unseen conversion of one type of eye into another. Upon this faith a humble shrimp imposes considerable strain. Moths, fireflies and Euphausiid shrimps, creatures all active in the dark, have special compound eyes which include a retina on which the multiple lenses focus at a common point to form an upright image. These shrimps, which seem to be, and are, classified as close 'cousins' to true shrimps, employ lens cylinders which smoothly bend the incoming light so that it all focuses at a common point, rather than forming multiple images as most compound eyes do. This feat of optical engineering has only been duplicated by humans in the last decade.

If this were not enough, Michael Land, a biologist from Sussex University, has observed that other shrimps have eyes which employ a different principle of physics, reflection from mirrors. The eyes have squared facets employed as radially arranged mirrors. It requires precise geometry to align such mirrors so that incoming rays are all reflected to focus at a common point, forming an image there. In an article entitled 'Nature as an Optical Engineer' Dr Land wrote:

I would guess that a refracting optical system, with refractive index crystalline cones, could not evolve into a reflecting system with squared multilayer-coated surfaces, nor vice versa. Both are successful and very sophisticated image-forming devices, but I cannot imagine an intermediate form that would work at all.[1]

No common ancestors or series, leading up to these two very different sorts of eyes in the same shrimp-like body, are known. Confronted with the evidence, I believe a reasonable Mr. Darwin would have opted for a theory of design. Over one hundred years ago he wrote: 'To suppose that the eye . . . could have been formed by natural selection, seems, I freely confess, absurd in the highest degree.'

Sound and Flight

Sight is supposed, although we cannot see it, to have evolved at least three separate times – in insects, squids and vertebrates. We are also invited to believe that flight, with all the critical structural alterations it entails, 'evolved' on at least four separate occasions – with insects, reptiles (pterodactyls etc.), birds and mammals.

Sound for flight in the dark: the bat, with its wings and sonar, is a perfectly aerodynamic mammal. To produce a bat from whatever its mammalian or reptilian ancestor was, must have involved innumerable transitional forms but none has ever been found. The oldest known bat is indistinguishable from modern bats. In spite of continuous fossil representation since the middle Eocene, the bats (Chiroptera) show no sign of evolving. An increase in the size of the skin-fold of such gliders as the colugo (flying lemur) is supposed to have led to the evolution of a bat-like wing. But sixty million years ago there were colugos like today's. There are about 2000 species of bat (mostly tropical), many with highly specialized organs. The following difficulties are encountered:

(a) How could the bones of the fore-limb, especially the digits, become gradually elongated? Any supposed intermediate structure between a fore-limb adapted for walking and one for flying would be of no use to the animal which would neither be able to walk, swim nor fly properly.

Natural selection would doubtless eliminate it.

(b) Structural modifications similar to those required by birds would have to occur, for example, light bones, enlarged sternum, growth of 'patagium' (bat's wing) and other essential skeletal and physiological changes like the articulation of wing-bone with shoulder-blade.[2] These, if the creature were to fly, would have to appear simultaneously.

(c) The pelvic girdle is rotated 180° in comparison with that of other vertebrates. How could reversal of such a bone complex as the pelvic girdle (although convenient for the suspension of the bat in caves) have been useful in its gradual stages, to any other creature? Strengthening of bat tendons must have occurred at the same time; furthermore, the knees bend backwards instead of forwards and it has a short sole and five toes of the same length, exactly the equipment needed to hang upside down.

(d) Specially designed milk teeth (with backwardly directed hooks) allow young bats to cling to the thickish hair on the underside of the mother's shoulder region. If both instinct and teeth to grip were not fully developed in bat progeny the consequences would be fatal for individual infants and therefore for bats in general.

(e) The mammalian diaphragm and rib-cage are modified in bats to support the action of wings in flight.

(f) The bat's detective mechanism is highly accurate; it is an echo-location system (sonar) of great complexity, in which high-frequency waves are transmitted through mouth or nostrils from a specialized larynx and the echoes picked up by large and specialized ears. The frequency of its supersonic squeaks is almost 70,000 cycles per second and it emits, in flight, up to 100 squeaks per second.

The bat distinguishes between its squeaks (which are coded differently in different species) and their echoes by the use of an ingenious mechanism: a small muscle in the outer-ear passage automatically contracts and closes the ear passage every time a squeak is emitted (at intermittent periods of as little as 1/200 of a second), so only the echo is detected.

This sonar is a marvellous discriminator: in a bat-swarm, in cave or night air, a bat can know its own sound among thousands of mobile neighbours, detecting its own signals even if they are 2000 times fainter than background noises. It can 'see' prey, such as a fruit-fly, up to 100 feet away by echo location and catch four or five in a second. And this whole auditory system weighs a fraction of a gram! Ounce for ounce, watt for watt, it is millions of times more efficient and more sensitive than the radars and sonars contrived by man.

The bat 'sees' with sound better than light. The idea that such an echo-location system (which would have to work straightaway or else accidents would eliminate the creatures) 'evolved' gradually by random mutation through unspecified 'ancestors' is inadequate. Indeed, that numerous changes must have had to occur simultaneously if the creatures were to operate effectively must prejudice the rational man towards creation theory.

A Precision Instrument of Aerospace

The origin of birds is largely a matter of deduction. If evolution from reptile to bird occurred, which it did (didn't it?), the theory insists that some kind of 'pro-avis' — a precursor of birds — is indicated.

The dinosaurs were a special group of reptiles that once flourished on earth as the mammals and birds flourish today. In Triassic rocks rest the bones of aquatic ichthyosaurs, plesiosaurs and nothosaurs, creatures that dominated the seas and rivers. In the air flew pterosaurs; on land a huge range of dinosaurs came and went over a period of 150 million years. Their graveyards are found on every continent and there are none alive today. Where and how did they originate? Why did they die out? We do not know.

Each distinct type of dinosaur appeared without transitional forms and, when they died out, left no obvious descendants. There were 'lizard-hipped' and 'bird-hipped' kinds, and evolutionists would say that their closest kin today were crocodiles and the birds. To the creationists, the dinosaurs were unique creations, now extinct.

Archosaurs, the most spectacular of the reptile groups which dominated the Mesozoic, include dinosaurs, pterosaurs and crocodiles. The pterosaurs are flying reptiles — *Rhamphorhynchus*, vulture-like *Quetzalcoathus* with a wingspan of well over thirty feet (three times that of an albatross) and *Pteranodon*. Just as dolphin-like ichthyosaurs and long-necked plesiosaurs were admirably engineered to fit their aquatic environment, to the extent that the eggs were hatched inside the female, so *Pteranodon*, with its twenty-five-foot wingspan and aerodynamic bony crest, illustrates brilliant aeronautical design. Like other Cretaceous fliers it probably glided, light as a kite, on thermal upcurrents and breezes found around cliffs. Only 30 percent of its structure is solid and sometimes the bone walls are only 1/25 inch thick. It was toothless and tail-less. With no known ancestors or descendants, *Pteranodon* was a distinct type.

If pterosaurs will not do, could 'pro-avis' have been a feathered reptile gliding down from trees? But there is no evidence that gliding precedes flight. As in the flying fish or flying fox, it is a distinct form of locomotion.

Some authors believe that birds evolved from small, agile creatures called coelurosaurs. Curiously, these bipedal creatures were 'lizard-hipped'; the bird-mimic *Ornithomimus,* which resembled an ostrich without feathers, is an example. Perhaps a coelurosaur developed a kind of feathery basket on each forearm with which to swat insects: perhaps it jumped higher and higher after the insects. The author of these ideas, John Ostrom, an expert on birds, has pointed out that the muscular motions for bagging flies and flying are quite different. He says: 'No fossil evidence exists of any pro-avis. It is a purely hypothetical pre-bird, but one that must have existed. . . .'[3]

There is no indication how arms turned into wings. When we find wings as fossils, they are completely developed and fully functional. *Ichthyornis* probably resembled the tern in its appearance and mode of life. A toothed jaw used in restoring the first specimen was later identified as belonging to an aquatic reptile, occurring in the same habitat, probably a baby

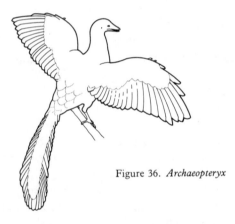

Figure 36. *Archaeopteryx*

mosasaur. *Ichthyornis* appears later than Jurassic *Archaeopteryx.*

Although no fossils lead up to or away from it, *Archaeopteryx* is often paraded as *the* link fossil. There are five specimens of this bird from Solnhofen limestone in Germany. Usually shown in textbooks is the Berlin specimen. It has birdlike features in the form of wings, beak, sclerotic eye-rings, fusion of the upper foot-bone into an extra section of the limb, an opposable hind-claw and, of course, feathers. Reptilian features include teeth in the bill, claws on the wings and a long bony feathered tail.

Are these features so reptilian? Just as *Pteranodon* is seen as a distinct, extinct type of reptile, so the creationist regards *Archaeopteryx* as a distinct, extinct type of bird. He argues that the 'reptilian features' fall within the sphere of variability of a bird. We ourselves have arm-bones similar to those of a bird, a whale and a bat, but are distinct from these types.

All birds have feathers: no other organisms do. *Archaeopteryx* has feathers. There exists absolutely no evidence for the evolution of feathers. The guess that DNA coding for scales 'must have' changed to produce feathers is entirely unsubstantiated. No intermediate scale-feather exists.

Feathers are aerodynamic beauties. They are light, the shaft being hollow, and quite different from the scales which are coded on to the feet alone of birds. They grow from capsules called 'pin feathers' and become lifeless when full-grown. A feather from wing or tail is composed of a shaft with branches, called barbs, arranged diagonally to the left and right. The barbs have branches to right and left called barbules. These overlap neighbouring barbules and are interlocked to each other by little hooks and eyelets.

Some large feathers contain over a million barbules, with hooks and eyelets to match, in perfect order. The feather is useless without this interlocking mechanism which acts something like an automatic zip fastener whose disturbance preening rearranges. When outstretched in flight, the hooks cause the whole wing-assembly to form a continuous sheet to catch the wind. The whole feather is a cohesive, elastic and light structure, well-designed to function as an air-resistant surface. Sensory receptors record its precise position. Over both wings they effect the continuous variations and fine adjustments of more than ten thousand tiny muscles attached to the bases of the feathers. Behold the parts of a precision instrument of aerospace, unparalleled in design and workmanship by human technology.

Feathers are in no way frayed or modified scales. They even arise from a different layer of skin cells. Whence evolved the pigment mechanism for colouring and patterning both plumage and egg? In the latter colours are laid down in the oviduct, in whose walls no pigment has been found. This indicates that the organization of pigment and pattern is coded into the avian DNA. By a megamutation?

In the earliest known fossils of pterodactyls, bats and winged insects the instrument of flight is fully developed. Just so, the wings and feathers of *Archaeopteryx* are as perfect as in modern birds. Asymmetric flight feathers resemble those of strong fliers; tail-feather arrangement parallels that found in modern swans and hens. It is a moot point just how good or bad at flying *Archaeopteryx* was. There are living flightless birds, such as the kiwi, with very small breastbones and not much of a keel (on to which the flight muscles are attached). Indeed, many birds 'have wings, won't fly'; these include emus, cassowaries, rheas, swimming birds (penguins), ostrichs, extinct dodos and moas.

As well as the feathers of a strong flier, *Archaeopteryx* has a robust furcula which has been interpreted as the site of origin of a well-developed pectoral muscle. The supracoracoideus muscle, the largest of those that effect the recovery stroke of a bird's wing, used to be thought essential for flight. Because *Archaeopteryx* lacked a bone necessary for its effective function in this way, it was considered flightless. Now it has been shown, by experiment, that it is not essential and that the dorsal elevators can alone effect the recovery stroke of the wing. These originate from the scapula which in *Archaeopteryx* forms an acute angle with the coracoid, as it does in adults of modern flying birds. In most flightless birds this acute angle is lost and the scapula is more nearly vertical. Despite interpretations to the contrary, there is 'nothing in the structure of the girdle of *Archaeopteryx* that would preclude its having been a powered flier.'[4]

You can see why it is argued that the *Archaeopteryx* falls within the sphere of variation of a bird. The bony tail? This is a distinctive feature and *Archaeopteryx* is, in fact, classified in a sub-class on its own. In the embryo some living birds have more tail vertebrae than 'Archy', which later fuse to become an upstanding bone called the pygostyle. It is certainly a permutation on the usual tail-end subroutine for birds, but so are its vertebrae, which have no saddle-shaped articulations. This 'reptilian' feature is also found in cormorants, darters, gulls and certain parrots.

The free (unfused) foot-bones and wrist-bones, found in *Archyopteryx*, are also found outside reptiles – in penguins. Indeed, *Archaeopteryx* had perching feet. To reduce weight, large bones in birds are hollow, strengthened with cross-struts inside. If the long bones of *Archaeopteryx* lack this characteristic, it is also missing in swallows, martins, snipe and canaries. Birds need an efficient respiratory system to serve the energetic demands of flight. A unique system of air sacs and capillaries, through which the flow is one way only, extends into some of the larger bones. The ventilation promotes not only the flight but, as air flows over resonating vocal chambers, song. Moreover, bird lungs have lung tubes not millions of tiny air sacs like those of reptiles and mammals. How could the transitional sac-tube organism have survived? Indeed, how could the whole, integrated 'flight-friendly' system have evolved piecemeal?

What about the teeth? No living birds have socketed teeth but some fossil ones did. Some reptiles have teeth, some have not. The same applies to fishes, amphibia and mammals. Following the analogy that toothless birds are more advanced, the toothless duck-billed platypus, or spiny anteater should be considered more advanced than humans.

In an interesting experiment[5] outer tissue was taken from the first and second gill arches of a five-day-old chick embryo and combined with inner embryonic tissue (mesenchyme) of a mouse, taken from the region where the first molar teeth form. Normally, the enamel layer of a tooth forms from the outer tissue and the underlying dentine and bone from the mesenchyme — if that tissue can interact with the outer tissue. The dentine can then induce the formation of a tooth.

Chick mesenchyme cannot form dentine so that its outer tissue never gets the chance to form a tooth — but in the experiment, where it was artificially exposed to the dentine-producing mesenchyme of mice embryos, it did. And it formed teeth! Teeth in a bird! This startling fact is explained by evolutionists as 'atavism', a doctrine of reversion: in this case the 'ancestral' genes for teeth are present, but suppressed by a mutation. A modification to the genetic programme for vertebrate mesenchyme has, in birds, disconnected it from the production of dentine and, therefore, teeth. If, as in the above-mentioned experiment, it can be reconnected, it will produce teeth.

In most modern birds, but not *Archaeopteryx*, the plan for the fibula and tibia leg-bones is modified, developmentally, so that the fibula is much reduced and the result is a single structure — the tibia with ankle-bones fused to it and the 'vestigial' fibula alongside it — which articulates with the foot-bones. Developmental manipulation of chick embryos by Frenchman Armand Hampé 'allowed' the fibula to attain the same length as the tibia — as it does normally in vertebrates; articulation with the ankle-bone changed accordingly.[6] Where the evolutionist sees Hampé's results as an expression of ancestral relationship in leg-bones, the creationist sees it as a modification, suitable for most birds, in the vertebrate programme.

A similar interpretation applies to wing-claws. In most modern birds they are suppressed but the young ostrich, rhea and the touraco of Africa have them. So do young South American hoatzin, a bird which shares a number of features with *Archaeopteryx*. It leaps, flaps and dives about the wooded rivers and swamps of the Amazon valley today.

Archaeopteryx could represent a group of distinct organisms that showed the characteristics of bird and reptile. No other fossils lead either to or from it. Stephen Gould and Niles Eldredge therefore claim it does not count as a missing link.[7] Links are not links if they are mosaics of complete functional traits from other groups. Whales and seals have a mixture of fish and mammal traits, penguins have fin-shaped wings and bats are a mixture of bird and mammal but no one calls them intermediate. No doubt *Archaeopteryx* is an odd mixture of subroutines but so are many other creatures.

Because bird types are found from the early Cenozoic, it seems only a matter of time before they are found in the Cretaceous or Jurassic beds. Already Cretaceous *Ichthyornis* shows signs of having been a tern: and in 1977 'Dinosaur Jim' Jensen found an avian femur and two connected shoulder bones in Jurassic rocks, where he had previously excavated his dinosaurs in Western Colorado.[8] The splendid isolation of *Archaeopteryx* was relieved by a bird which predated it.

There is no decisive genetic or fossil evidence for evolution from scale to feather, cold to warm-bloodedness, non-flight to flight. Almost every tissue, bone and organ differs dramatically in birds. Eggs have now to be incubated at constant temperatures. Birds have a third eyelid, a 'nictitating' membrane which functions as a transparent windscreen wiper during flight. It is drawn protectively across from the inner angle of the eye hundreds of times a minute, cleansing the surface of dust, small flying insects and the other debris which rapid flight encounters. A fast-moving vehicle needs uninterrupted vision. Man and other mammals possess a rudimentary nictitating membrane in the inner angle of the eye and this is taken by Darwinians as proof of common ancestry. But some sharks, reptiles, rabbits, the walrus and the duck-billed platypus have such a membrane. No one claims any special relationship is thereby proved – unless the inference of thematic vertebrate design is drawn. Perhaps all vertebrates possess the full basic plan from which 'seed-information' is genetically expressed or suppressed. Appropriate gene control would have been initially encoded into different types of vertebrate. In this case, the upshot would be an operative nictitating membrane (or other feature) where it was required, and an apparently vestigial or adapted appearance in other types of vertebrate.

Birds have air sacs for buoyancy, bills, powerful breast muscles, avian feet, gizzards instead of large jaws and heavy teeth for mastication, a large strong heart, wings, feathers and other non-reptilian features. Such structural discontinuities isolate the avian sphere of variation from that of reptiles or mammals. Such features, which add up to a distinct stock, are all essential to their poetry in motion, to their lordship of the air.

Helicopter in a Thimble

Insects are the only invertebrates that possess the power of flight. This enables them to inhabit a great variety of environments; they are the most diverse of all animals with approximately one million species described.

Typical insect wings are flying machines no less efficient than those of bats, birds or Boeings. They develop in pairs from the thorax and consist of a thin membrane stiffened by numerous veins, resulting in a functional compromise between weight and strength. In the body are air channels used for respiration; air diffuses along these channels to the interior parts of the insect. Evolutionists sometimes claim that the veins have a consistency of pattern from one insect to another, indicating a common ancestry. For the creationist, this consistency indicates the repetitive, thematic presence of a good mechanical design; it no more indicates common ancestry than does DNA.

The front portion of the wing is stiffened with a heavy vein, and the wing becomes thinner and more flexible toward the trailing edge. This semi-flexible structure is capable of very strong sculling action. Insects with two pairs of wings sometimes have them joined by means of hooks and grooves to create a single sculling unit (bees, ants, wasps and many butterflies and moths). Dragonflies do not have them joined and overcome the problem of air turbulence by beating front and rear pairs alternately.

Some insects, such as flies and beetles, only have one pair of wings. The forewings become the hard elytron (wing-cover) of a beetle; or the hind-pair have been modified into tiny balance organs called halteres, that move up and down like the wings but in opposite phase.

Winged insects take advantage of the same aerodynamic principles as manmade aircraft. At each flap the wing blade is twisted from elevation to depression. In one complete cycle the wing tips describe a figure-eight pattern. Many insects can hover or fly backwards by changing the angle of the figure eight. Some very good fliers, such as flies, bees, wasps, ants and some butterflies and moths can fly sideways or rotate the head about the tail by employing unequal wing movement. You can fit nature's helicopters in a thimble.

Wing movement is so efficient that it produces a polarized flow of air from front to rear, during 85 per cent of the wing-beat cycle. Insects with large wing area, such as butterflies and dragonflies, have the wing muscles attached directly to the wings and one nerve impulse creates one wing-muscle contraction. Butterflies have a wing-beat of about ten per second. Bees, wasps, ants and flies must combine excellent flying ability with small wing area. They compensate with a very rapid wing-beat (try flapping your arms even three times a second!) whose frequencies vary from fifty-five per second in some beetles to 200 in bees, 500 in whining mosquitoes and an incredible 1040 per second in the midge.

Nerve tissue is incapable of firing this many times per second, so that a mechanism is employed whereby a single motor-nerve stimulus can cause several wing-beats, in a manner analogous to mechanical oscillations produced by flicking a ruler in a desk-lid. In the insect's case the fulcrum is a portion of the thorax. The natural elasticity of the thorax in flies and some beetles enhances this oscillation by imparting a 'click' action in which the wings, as they pass the centre of the wing-beat action, are driven swiftly to the extremes by the spring action.

Insect wings appear in the fossil record, from the first, fully formed. No evolution, no intermediate forms are found, any more than transitional forms between orders of insects, such as grasshoppers, bees or damsel flies, are found. Insects have always been numerous and varied; they retain as many types in the present. They do not demonstrate plasticity. Indeed, it strikes me as strange that some ancient stock (say fishes, amphibians) should have been thought to split into a different group with astonishing plasticity, and worked up to human status, while their brothers have remained, with almost total rigidity, as they always were. These persistent types, of which insects are one of many, are a thorn in the side for evolution.

Amber fossils indicate that ancient wings are essentially identical to modern forms. An extinct form of dragonfly found in coal (Carboniferous) deposits had wings with a seventy-centimetre span. No intermediate forms exist between this type of wing, which cannot be folded, and those whose different mechanism enables them to be flexed and folded back into a resting position.

There is no clue from the fossil record as to the origin of the iridescent wings of a dragonfly or the flight of the bee. A leap of faith is needed to believe in the evolution of an eye, a feather, an insect wing or any other special organ.

Notes

1   Land, M., 'Nature as an Optical Engineer', *New Scientist*, vol. 84, no. 1175, October 1979, p. 13.

2   McMurray, N., 'Chiroptera', *Creation Science Movement Bulletin*, no. 158, 1968, p. 1.

3   Ostrom, J.H., 'Bird Flight: How Did it Begin?', *American Scientist*, January-February 1979.

4   Feduccia, A., and Olson, S., 'Flight Capability and the Flight Girdle of *Archaeopteryx*', *Nature*, vol. 278, 15 March 1979, pp. 247–8; Feduccia, A., 'Feathers of *Archaeopteryx*: Asymmetric Vanes Indicate Aerodynamic Function', *Science*, vol. 203, no. 4384, 9 March 1979, p. 1021.

5   Kollar, E.J., and Fisher, C., 'Tooth Induction in Chick Epithelium: Expression of Quiescent Genes for Enamel Synthesis', *Science*, vol. 207, 1980, pp. 993–5.

6   Gould, S.J., *Hen's Teeth and Horse's Toes*, Norton, 1983, pp. 184–5.

7   Gould, S.J., and Eldredge, N., *Paleobiology*, vol. 3, 1977, p. 147.

8   *Science News*, vol. 112, 24 September 1977, p. 198; Ostrom, J.H., *National Geographic Magazine*, vol. 154, no. 2, August 1978, p. 152.

16 Time and Space

Who's Afraid of the Big Bang?

We are here. The universe is here. How? There are three main possibilities.

(a) The universe always existed.

(b) The universe sprang into existence when, at the beginning of time, nothing nowhere for no reason exploded (Big Bang Theory). Attempts have been made to reconcile (a) and (b), for example, postulating an 'oscillating universe' which disappears in a 'black hole' and reappears as a 'white hole'.

(c) From mind to molecules; mind preceded matter and created the cosmic drama which it now sustains according to a recognizably lawful programme.

Figure 37.

All three are natural explanations. Of course, if there exists a philosophical equation of 'nature' with physico-chemical operations alone, then (c) is excluded. But just as soul animates the body, so the adherents of (c) believe the universe is the 'body' of an Animator. What, at the heart of things, is more natural than this Power?

The laws of gravity, of motion, of thermodynamics (heat energy); the conservation principles, for example, conservation of mass, energy, momentum, electric charge; and all other basic laws are presumed to have

229

remained, through all time and space, unchanged. The laws of the universe are not evolving. Nevertheless, the neo-Darwinian idea of evolution seems to require an undiscovered principle of increasing organization, which introduces new systems into nature and develops existing systems into higher ones. Conversely, for the creationist, creation is finished and he would expect its basic laws to be conserved.

The First Law of Thermodynamics (the law of energy conservation) states that energy, and its correlate matter, is neither being created nor destroyed. In other words, an initial input of power now exists unaltered in quantity, as the flux perceived as motion in stars, galaxies, atoms, etc. The inference here is that there is nothing in the present structure of natural law which can account for its own origin.

'Motion', observed Sir Isaac Newton, 'is much more apt to be lost than got, and is always upon the decay.'

The Second Law of Thermodynamics (the law of energy decay) states that, with time, all systems tend, unless there is an external input of energy, to run down. For example, the paraffin wax in a candle is composed of hydrocarbon molecules which, by virtue of their structure, possess much chemical energy. When lit, the candle will burn and chemical energy will be converted into light and heat energy. The candle burns spontaneously, but it will never 'unburn' itself. All spontaneous processes tend to change order into disorder and organized energy into random heat energy. The entropy of a system is a measure of its degree of disorder, and we expect to find a general increase in entropy.

Jeremy Rifkin, an evolutionist, has written: 'The entropy law will preside as the ruling paradigm over the next period of history. Albert Einstein said that it is the premier law of all science, Sir Arthur Eddington referred to it as the supreme metaphysical law of the entire universe.'[1]

What meaning has this law for the question of origins? Firstly, if the cosmos were infinitely old we would expect to find it had completely run down. Unwound, it would have suffered a 'heat-death'. That it has not implies that it is not infinitely old and therefore must have had a beginning. If the first law precludes the cosmos from having started itself (possibility (b)), we are led to the conclusion that an outside, non-material power generated it.

The second law, implying that the universe had a beginning, precludes the possibility of infinite, eternal matter (possibility (a)). The entropy law and all related consequences, including time, must have been suspended at creation. In other words, the universe was wound up in no time. It is no less scientific to postulate an 'infinite moment' of creation than such a moment of destruction in a 'black hole'. Within such singularities, we are informed, the known laws of physics are suspended, inoperant or overruled.

Wound up in no time; unwinding in time. The creation of energy and its correlate, matter, is a difficult idea to grasp. Our minds are programmed to operate, not in a timeless moment of creation, but in time, as the universe runs down. Creation 'goes against the grain'. It is possible that it is, in the last analysis, beyond human comprehension. Nevertheless, theoretical physicists work up various mathematical answers. One explanation is that creation was born of a structureless 'cosmic egg', an incredibly 'big bang'; another indicates an instant start, of maximum order, after which increasing disorder would 'take time'.

Possibility (c) remains inviolate by the laws of thermodynamics. It predicts, amid increasing disorder, a conservation of cosmic laws. Of course, this does not 'prove' a Creative Power; and materialistic ideas of anti-material and multidimensional universes have been used to 'plug the gap'. Electron and positron, proton and anti-proton, positive and negative, black and white: have you considered that the antipole of matter might be intelligence or, of a creation, its Creator? Intelligence *is* a strong creative power. Great intelligence perceives and creates great order.

Parapsychologists study extra-sensory perception. Alongside physicists they seem to show that physics, although a powerful body of knowledge, is incomplete. Science has neither proved nor disproved the serious possibility of intelligent creative power outside the material dimension. Whether as human (locked to a brain) or cosmic mind, this 'thought-world' would exist as the basis and creator of space and time — the matter and energies we call the 'real world'. Insufficient knowledge or imagination does not exclude the possibility of its presence nor, with the right instrument and technique, its fuller discovery. Some people claim that mind, properly attuned by correct yoga or meditation, may constitute just such an instrument.

This brings us to a second point. Statistical thermodynamics shows that the organized complexity (order) of a structured system tends to become disordered. A correlation of this is that the information conveyed by a communicating system tends to become distorted and incomplete — you might say mutant.

It is worth remarking that Clausius, Carnot and Kelvin were working out their thermodynamic generalizations on steam engines in the period immediately after *The Origin of Species* was published. Darwin was unaware of the impact of their labours on his theory.

Life is Unnatural

The arrow of time points downwards. Just as water finds its own level, so chemical reactions tend toward the stability of equilibrium and a 'heat-death' of the universe, where everything is at the same temperature with no energy available for use.

Life is different. Death levels individuals but, like a spring, generation after generation, life itself wells up. It keeps juggling; it stays unstable. Its molecules are built up into improbable, unnaturally complex groups. Evolution requires an increase, not just in physical size of molecules but in organization. Organized systems are purposive, assembled element by element according to an external 'wiring diagram', with a high information content. You may throw letters together and, by chance, make words. But organization requires that words are assembled more often than chance alone would achieve, and grouped in sentences to make meaningful statements.

Wherever such high-grade organization appears to counteract, at least temporarily, the general tendency to disorder, dilapidation and 'deadly' randomness, inspection reveals a feature behind it − plan. A pile of bricks and planks becomes a house, a seed grows into a tree. In both we see plan, one in the builder's mind, the other encoded in DNA.

'Hold on!' cries the evolutionist, 'a plan represents an increase in information. You say the Second Law of Thermodynamics forbids this happening spontaneously. That may apply to *closed* systems. But the earth is an open system with an external source of energy (the sun). This energy is sufficient to generate local increases in information, such as complex chemical compounds and, eventually, life-forms.'

'Untrue,' says the creationist. 'Raw, uncontrolled energy is destructive. To build the biologically complex from the simple (and a cell is certainly very complex) four, not two, conditions are required:
(1) An open system (such as earth)
(2) An adequate energy supply
(3) Energy-conversion mechanisms
(4) A control system directing, maintaining and reproducing the energy-conversion systems.

The hypothetical primordial earth would have satisfied only (1) and (2). Yet (3) and (4) are essential criteria for the development and maintenance of biotechnology. For life (3) means photosynthetic, respiratory and other metabolic systems; (4) means reproductive apparatus including DNA (the genetic code). How did the precise, highly informed engineering required for (3) and (4) occur by chance?'

Time is no help. Biomolecules outside a living system tend to degrade with time, not build up. In most cases, a few days is all they would last. Time decomposes complex systems. If a large 'word' (a protein) or even a paragraph is generated by chance, time will operate to degrade it. The more time you allow, the *less* chance there is that fragmentary 'sense' will survive the chemical maelstrom of matter.

And what sense is there in life? Why is it that a highly complex and diversified factory such as a cell should, with all its strength and purpose, submit to the command 'Live!' Whence issues this command? What is the use of survival? In the end survival comes down to existence and what is existence for? Chemicals alone, unless they are most unnatural ones, can issue no such command. It is issued only through the complex machinery of protoplasm. How simple, how fragmentary can any machine become before it ceases to operate? Fragments, in the unlikely event of sufficient being present at any one time or place, are under no compulsion to assemble themselves into a factory-like bacterium. The reverse is more likely, as the physical laws inexorably command their destruction into smaller, stabler entities. In this sense life, like any purposeful machine, goes right against the grain. It is unnatural.

Life the Unlikely

Is a vortex of water gurgling down a plughole a clue to the origin of life on earth? A vortex is a 'dissipative structure', i.e., one maintained by a continuous flux of energy. All living things, consuming food for energy metabolism, are in this sense dissipative structures. In the 1970s a mathematician, Ilya Prigogine, proposed that 'fluctuations' in dissipative structures could generate higher forms of order – vortices upon different-shaped vortices.

The problem for evolutionists is to explain the origin of the high level of organization – comparable to writing as opposed to ripple-marks on a beach – which is found in dynamic, biological compositions. So severe is the problem that a number of scientists grasped at Prigogine's Nobel-prizewinning idea as a solution. But Prigogine himself acknowledges that there is no evidence at all that life began in this way. He notes that the probability of life, with its large molecules, just assembling itself at ordinary temperatures – or any other kind of temperatures – is vanishingly small.

It does not matter how many 'prebiotic attempts' there were. All calculations, by creationist and evolutionist alike, show how unlikely life is.[2] Normally, if an event is one in a thousand, scientists consider it statistically negligible. The odds against the formation of life are billions of times greater — yet the cosmic casino turned up trumps. Life, which probably began as a unique event, is here! However 'impossibly', it must have assembled itself!

Or must it? Although it is impossible to prove an 'impossibility', the creationist believes not. No Canute, he is not philosophically obliged to stem the statistical tide against chemical evolution. Rather, knowing that technological inventions are all 'unlikely' but that intelligence works wonders, he runs with it. 'Cosmos' is a Greek word meaning 'order' or 'arrangement'. The cosmos is power. The joyous cosmology of the creationist celebrates the biological arrangement of power.

Rock of Ages: Ages of Rock

If natural selection and mutations are insufficient agents of evolution and the basic laws of physics are inimical to it, it is proper to consider another approach to the question of origins. If time is unhelpful, upon what can evolution depend?

Is not the fossil record cast in ancient rocks? For all its shortcomings, surely it represents to the reasonable man a powerful evidence that evolution must have occurred? This is the point to mention that by no means all creationists believe that creation myths, whether Biblical (from Mesopotamia), Polynesian, Scandinavian or whatever, are other than myths. They do not take simple metaphors as literal truth, any more than the plum pudding or planetary models of the atom are considered final. Nor, on the other hand, do they rule out intelligent codification of life-forms.

Modern geology is based on two major premises — that the earth has great age, and that Hutton's principle of uniformitarianism (summarized in the aphorism 'the present is the key to the past') holds good.

The stratigraphic column is built up of rock units, for example, coal and chalk measures, and subdivided into beds and bedding planes. By using certain fossils as indicators, the column can be divided into time zones. This 'biostratigraphic column' is defined by zone or index fossils, which occur in certain strata. For instance, any rock containing fossils of one type of trilobite (*Paradoxides*) is called a Cambrian rock: of another type (*Bathyurus*) Ordovician. In this way a relative geological timetable is compiled.

Absolute dates, for example, 430–500 million years ago for the Ordovician period, are ascribed by using igneous rocks for calibration points. These occur among the sediments in the form of sills and dykes, and radiometric methods can be used to date them accurately. Sediments on which the dated igneous rocks lie must predate them and so be older: those overlying the igneous rocks must be younger. So the geologist ascribes ages to all his strata.

No geologist claims his understanding is flawless. While evolution requires long periods of time (and is the habitual mode of geological thinking), the creation of a series of machines may occur 'all at once' or be extended successively over a period of time. Many creationists, while conceding that the question of the age of the earth and the universe is an important one, believe that it is independent of the question of whether or not creation occurred. They insist, for reasons explained above, that time is not the critical factor and are prepared to allow whatever shifting time-scales are fashionable. These have changed, doubling on average every fifteen years, from about four million years in Lord Kelvin's day to 4500 million now.

Both poet and atomic physicist will agree that metaphors are not to be taken literally. The book of Genesis may express the truth of creation in a simple, allegorical way – nevertheless Biblical literalists challenge orthodox geology in three ways. Firstly, it was noticed that geology relies heavily on the evolutionary interpretation of the fossil record because the relative ages of various rock formations are determined largely in accordance with the presumed stage-of-evolution of the fossils in them. 'The only chronometric scale applicable in geologic history for the stratigraphic classification of rocks and for dating geological events is furnished by the fossils. Where there appears to be a discordance between physical and fossil evidence as to the age of any series of beds, fossil evidence is usually preferred – even to radiometric dating. There is no one area where the whole series of fossils is found; but an evolutionist, having decided that living things evolved from less to more complex, arranges finds from different areas on the assumption that his decision is correct.

Herein lies a powerful tautology, a circular argument. The assumption of evolution is the basis upon which index fossils are used to date the rocks; and these same fossils are supposed to provide the main evidence for evolution. The fossil record, itself based on the assumption of evolution, is interpreted to teach evolution. By this sort of reckoning, the main evidence for evolution is the assumption of evolution.

We have already seen several 'index' fossils that, after an absence of millions of years, have turned up alive. Examples are the tuatara lizard, the small mollusc *Neopilina gàlatea*, the maidenhair tree and the dawn redwood; most celebrated is the 'link fossil' coelacanth fished up off the coast of Madagascar in 1938 after seventy million years' absence. Any rock dated according to coelacanth index is not now seventy million years old, but anywhere between a hundred years and seventy million. Because most index fossils are small marine organisms, for example, molluscs, and the ocean depths are mostly unexplored, it is possible more index fossils will be discovered.

But even if not, it is of little consequence to the argument. Evolution has to explain not only persistent species but persistent kinds. As regards classification, all kingdoms and subkingdoms are represented from the Cambrian onward. All classes of the animal kingdom are represented from the Cambrian onward except insects (Devonian onward) and (perhaps) vertebrates and moss-corals from the Ordovician. All phyla in the plant kingdom are represented from the Triassic onward except bacteria, algae, fungi (pre-Cambrian onward); mosses and horsetails (Silurian onward); diatoms (Jurassic onward) and flowering plants (Cretaceous onward). As we noted earlier, orders and families (as well as kingdoms, phyla and classes) appear suddenly in the fossil record, with no indication of transitional forms from earlier types. This is the case even for most genera and species. Index-fossil sequences such as micraster, ammonites and trilobites indicate only variation ('micro-evolution').

Secondly, fundamentalists interpret geology in terms of a young earth. A global catastrophe, resulting in a flood, was followed by several thousand years of uniformitarian processes. For non-gradualists (catastrophists) the geologic systems represent broad ecologic and sedimentary zones. They include fossils of creatures that lived at different places at the same time. The geologic column therefore represents the usual sequence in which rising, turbulent flood waters buried plants and animals from these different zones. Deposition and flood currents would not be entirely predictable but normally soft-bodied bacteria and worms, that inhabit poorly oxygenated depths below the sea floor would be discovered in the lowest strata. Shelly forms from the sea floor would be found next, followed by marine organisms such as fish. Later would come land-plants and creatures; those best able to flee the waters would be last and least likely to be buried by rapid deposition of sediment and fossilized (see fig. 38).

R. V. Gentry has found some 'haloes' which he believes indicate, if not a divine creation, at least a young earth.[3] Some radioactive isotopes of polonium, bismuth and lead have half-lives of only a few minutes. This means, effectively, that they swiftly decay out of existence. Inclusions of their decay 'haloes' in pre-Cambrian granites and coalified wood show as dark circles. As they would not appear in molten rock, the indication is that both granite and haloes were formed at the same time and, within minutes, radioactivity darkened the rock crystals. Dr Gentry, a meticulous scientist who has published work in leading scientific journals, has for fifteen years

Standard System	Corresponding stage of the flood.
Recent	Period of post-Flood development of modern world.
Pleistocene	Post-Flood effects of glaciation and pluviation, along with lessening volcanism and tectonism.
Tertiary	Final phases of the Flood, along with initial phases of the post-Flood readjustments.
Mesozoic	Intermediate phases of the Flood, with mixtures of continental and marine deposits. Post-Flood possibly in some cases.
Paleozoic	Deep-sea and shelf deposits formed in the early phases of the Flood, mostly in the ocean.
Proterozoic	Initial sedimentary deposits of the early phases of the Flood.
Archaeozoic	Origin of crust dating from the Creation Period, through disturbed and metamorphosed by the thermal and tectonic changes during the Cataclysm.

Figure 38. *Fundamentalist chart describing the ages of the earth* by H. Morris in *Scientific Creationism*, p. 129

been challenging the 'big-bang' cosmology. The world's greatest geochronologists have been unable to refute his work.

The third challenge attacks the validity of radiometric dating.

Using the three challenges in their interpretation of the known data, 'young-earth' creationists have attempted to assemble a geology based on non-evolutionary premises. No less than the evolutionist model, theirs has problems. For example, why do we find fossil successions? Oyster-like creatures are found from bottom to top of the record – strange for slow-moving bottom-dwellers. This kind of succession is particularly marked in chalk deposits where a definite succession of different species of the same type of creature are found, separate and unmixed, at different levels. Chalk cliffs are not easily explained by young earthers. Nor are the layers of salt hundreds of feet thick thought to have been laid down as seas slowly evaporated.

If they all once lived together, why do whales, seals, placoderms or ichthyosaurs not appear with modern fishes in marine Devonian environments? Why should large animals not settle at the bottom and smaller ones at the top of sedimentary layers? Why, if the flood took place rapidly, are sandstones nearly always void of fossils? Uniformitarians reasonably explain that, over a period, shells are oxidized and abraded out of existence by the sand – but is 10,000 years enough? How do young earthists explain mud-cracks found lithified and buried between two layers. Mud-cracks can only form when the surface of the mud dries out – in a flood? Similarly, in a flood the large stones found in conglomerates would sink: yet they are often found, separated by a clean, sharp line, on top of deep beds of fine-grained sediments. The line indicates that these sediments must have hardened – in a flood? – before the conglomerates were dumped. Hard, worn-round conglomerate boulders may be laden with fossils which the flood was supposed to have caused.

Indeed, if you count all the fossils, including coal from plants and oil from plankton, the earth could not have supported such life in a single generation. Nor could turbulant floods have laid it down in clearly defined rock strata. This all strongly indicates an ancient earth.

You can see how, if man's sin triggers a nuclear holocaust, the prophecy of a second destruction by fire might occur. But how could man's sin trigger Noah's flood? 'Young earthists' will complain I have skimped their case. They must, therefore, summarize and collect their most powerful evidence for the catastrophist picture. They will not amuse orthodox geology, which has laboured for 150 years and reversed that picture.

How central to the origin's debate is the question of age? The idea of *biotechnological design*, in modern terms, is a rational explanation of the scientifically-discovered facts. Certainly, evolution's pillar of faith is established on the ages of rocks but for a creationist age and design are distinct issues: deliberate design, not age is the crux of the matter. A creationist is not obliged to embrace Old Testament fundamentalism; indeed, many believe the creation of life may have happened in stages, each generating life-forms more complex than the one before.

After the cosmos, suns and planetary systems (including earth) were formed, bacteria and photosynthetic blue-green algae would create and maintain conditions fit for life. On this foundation a biological pyramid, a hierarchy of bodies could arise. At each new level an influx of plant and animal archetypes would become established through thousands or millions of years, until the wave of the next stage overtook it. Could mankind be the product of a recent stage and represent the crown of creation this far? At any rate, for the modern creationist the Rock of Ages, not the ages of rocks, is important.

Notes

1   Rifkin, J., *Entropy, a New World View*, Viking Press, N. Y., 1980, p. 6.
2   See Kaplan, M., ed., *Mathematical Challenges to the Neo-Darwinian Interpretation of Evolution*, Wistar Inst., Philadelphia, 1967.
3   Gentry, R. V., *American Journal of Physics*, vol. 33, 1965, p. 878; also *Nature*, vol. 252, no. 5484, pp. 564–6.

17 Adam and Evolution

In Hebrew, Turkish, Hindi and several other languages, Adam means 'man'. Eve is from 'Hawa', a Hebrew proper name. Its meaning is less clear but may be connected with the Hebrew word for 'life'. Eden, their home, in Sumerian means plain. The septuagint translates Eden into the Greek word *paradeisos*. This word is of Persian origin and means a watered pleasure-ground or hunting-park, of the kind that Xerxes and Cyrus devised. Need we look further back for our origin than paradise?

How dry old bones kick up the dust!

Perhaps more than any other science human prehistory is a highly personalized pursuit, the whole atmosphere reverberating with the repeated collisions of over-sized egos. The reasons are not difficult to discover. For a start the topic under scrutiny − human origins − is highly emotional, and there are reputations to be made and public acclaim to be savoured for people who unearth ever older putative human ancestors. But the major problem has been the pitifully small number of hominid fossils on which prehistorians exercise their imaginative talents. . . .[1]

So wrote Roger Lewin in an article about the opening of the International Louis Leakey Memorial Institute for African Prehistory (TILLMIAP) in Nairobi. The trouble is simple enough: there is not much material for anyone to work on. As we have seen, animals consist mostly of fragments of jaws, broken skull pieces or feet. No complete or even half-complete skeleton linking man with the rest of animals has ever been found. But the more fragmentary the remains, the more sweeping the claim, the more presumptuous the presentation of the evidence, and the brighter the blaze of publicity.

The number of experts and the investment in work and reputations has grown enormously since the days of Dubois, von Koeningswald and the other pioneers. But is the situation clearer? Have 'missing links' really been found? There are as many 'ape-to-man' genealogies as experts, and there was no shortage of either during the long years the Piltdown fraud remained undiscovered. Today's research may be more honest, yet it is even more strongly motivated by a wish to validate evolution, to 'see' evolution proven in the fossil finds. In his book, *Origins*, Richard Leakey refers to '. . . the Leakey tradition, that you look and look again until you find what you know must be there'.[2] It is difficult to imagine a more dangerous philosophy for an impartial scientific worker to adopt. Many evolutionists share it, though perhaps they are not all clear enough in their minds to say so. If there is just one thing that every evolutionist agrees on, it is that evolution happened: it's just a matter of looking until you find the evidence.

Are You Above it?

Chimpanzees, gibbons, orang-utans and gorillas are pongids. Human and 'ape-men' are homin*ids*. Homin*oids* include both pongids and hominids: it is within the primate super-family, Hominoidea, that evolutionists search for human origins.

The Fuegian Indians that Darwin visited were not pongids. They were a people who lived, naked, among the rocks of ice, wind and fire in Tierra del Fuego, at the southern tip of South America. They numbered about 7000 in Darwin's time: by 1932 there were only forty-two survivors, including half-breeds, and now they have probably disappeared. Darwin's opinion of them gave his Victorian contemporaries a comfortable feeling of their own supremacy.

I could not have believed how wide was the difference between savage and civilised man; it is greater than between wild and domesticated animals, inasmuch as in man there is a greater power of improvement . . . the difference between a Tierra del Fuegian and a European is greater than between a Tierra del Fuegian and a beast.[3]

Three Fuegians were captured and brought to England for education. They stayed for a few years and were returned. Darwin talked with them on board ship. He found them quite human in their sympathies but scarcely articulate. Captain Cook compared their language to a man clearing his throat and Darwin came to the conclusion that it consisted of only about a hundred different sounds. Many animals make a dozen or more different sounds, so that comparison with advanced animals seemed apt enough.

Of course, man's body is animal but he has gifts which distinguish him.

Certainly, in later life, Charles Darwin was so astonished at the apparent 'rehabilitation' wrought by missionaries on natives that he contributed regularly to the funds of a missionary society. But, although Christianity might help the most degraded savages evolve into more manlike specimens, it did not change Darwin's philosophy about them. In 1921–3, however, when the traditions of the tribe were still intact, two Austrian priests who had been trained in anthropology went to live among the Fuegians. They found the tribesmen set a high standard of morality and ethics; they believed in a Supreme Being who had created the world and the framework of society and prayed to Him in need, especially at death.

An English missionary, Thomas Bridges, made Tierra del Fuego his home and brought his family up there. He found the natives moral, kind and sociable. They had respect for family life and were not cannibals. Mr Bridges spoke the language of the tribe. He compiled a dictionary that was not exhaustive but contained 32,000 words and inflexions. The vocabulary was rich and the grammatical constructions involved. This was the language they spoke from an early age, and throughout life.

Darwin was wrong about the Fuegians. Myopic, due to the imposition of his own world-view, or blinkered because he could not communicate with them, he misjudged them. They were true men as much as he was, with full intellectual faculties and spiritual qualities that they did not show to casual passers-by. If Darwin could misjudge a living 'specimen' with whom he could communicate, what trammels are there to restrain the speculation of an ardent evolutionist (and his artist associate) over a few broken bones?

Essential Differences Between Man and Ape

(a) Brain volume
(b) Hands and feet
(c) Pelvis and gait
(d) Teeth
(e) Face and jaw
(f) Language, literacy and numeracy

To see a troupe of langurs hopping across the road from one belt of trees to another in the misty Himalayan foothills is to know that primates and man are akin. Closer still, the hominoids form a group of similar or homologous types. It is no surprise that apes should, on the basis of thematic design, resemble man in most features of anatomy, physiology and biochemistry. In respect of some proteins for which sequence data has been run, the likeness between chimpanzee and man is very close. But a man is a man, a chimp is a chimp, for all that.

The basic *brain volumes* of man vary from 750–2350 ccs except for microcephalic idiots of 500 ccs or less. There exists wide variation in a normal population but the adult male norm is about 1500 ccs. Ape skulls vary from about 90–685 ccs and some possess a calvarium or bony crest, which man does not. It is not so much brain size *per se* which is important. A small watch works as well as a large clock and a mouse is as clever as a rhinoceros and a woman no less intelligent than a man. What appears to count is the ratio of brain to body size. Where it suits, cranial capacity is taken as an indication of intelligence and skulls are arranged in ascending order of size. But it is another story in the case of Neanderthal man, whose cranial capacity was, on average, equal to or greater than that of modern man. Indeed, the large size of man's brain in 'savages' was one reason why Wallace, Darwin's partner in 1859, later distanced himself from Darwinism.

It may be significant that man has legs longer than his arms; long arms in apes are associated with 'brachiation', i.e., swinging through the trees. Man's torso is shorter than the apes. His posture is upright with hands free while walking; although apes sometimes walk erect they mostly 'knuckle-walk'.

Man's specialized foot for walking on is fully expressed even in the embryo. It is arched both across and lengthwise but cannot grasp objects. In fact, the foot of a mountain gorilla from Zaire differs from the hand-like foot of other apes, resembling that of a human. Its arms are not too long, nor its legs too short – a young gorilla can rear up and walk in a human way, resting on the sole of its foot rather than the side. But chimps and lowland gorillas have typical hand-like feet, so that the origin of the 'human' foot is unclear.

How provident for the hand that *Eusthenopteron* (chapter 13) happened to have two bones instead of one in part of its fin. Did they evolve into the radius and ulna, which form an ingenious joint permitting rotation of the hand around its long axis? Or was this facility, basic to skilful manipulation, part of the plan for man? An opposable thumb enables him to manufacture as well as use tools; and the area of his cortex devoted to hand movements is relatively much larger than that found in apes. While the ape cannot move its fourth and fifth digits independently, man manipulates fire, machinery, electricity and, now, nuclear power. Evolution has not trumped creation with a suite of intermediate ape-to-man hands.

The ape's pelvis is not designed for the upright position. Man's hip-bone is short and broad, the ape's long and narrow. This difference in shape is related to the curvature of the backbone which, in man, sweeps in an extended S to facilitate the upright posture.

Man has small teeth and short eye-teeth (canines): apes have large teeth for masticating raw food and large canines with which they fight. Man eats meat, apes generally do not. It is purely speculative why an ape should have modified a successful dietary habit to scavenge or hunt. Only theory requires it. An ape's lips are thinner and more mobile. Man's nose is more prominent and his nostrils less flaring. His jaw is parabolic rather than U-shaped as in most apes. And, an interesting point, the ape jaw is projecting (prognathous).

Why interesting? Neoteny is a process whereby juvenile traits of ancestors are, it is supposed, retained into the adult stages of development in the more 'advanced' kind. For example, newborn and juvenile chimps exhibit skulls considerably more rounded and 'human' in shape than adult chimps. In other words, the juvenile skulls more nearly resemble those of modern man than they do those of their own adult forms. It is possible to parade sequences of skulls which illustrate this fact.

As A. Montagu wrote: 'In surveying the development of the concept of neoteny it becomes clear that the morphological changes in the varieties of humankind have been mainly brought about by the retention into adult life of traits principally characteristic of the foetus.'[4]

In a nutshell, it is supposed that modern humans evolved from apes by retaining juvenile traits. It could also be argued, although not by an evolutionist, that the similarity between adult human and juvenile ape skulls indicates that, in this case, the human, not the apish, type is basic. In other words, that recapitulation by acceleration occurred in the ape, so that pongids radiated from a human ancestor.

A Finnish lecturer in palaeontology has fuelled the debate. For him both contemporary and fossil forms disprove the standard notion that man is descended from the apes. It is well recognized that man is a generalized (though advanced) organism. Inhabiting a body capable of many activities, he is very adaptable and can make himself at home almost anywhere. Kurten points out that specialized dentition in apes (such as their premolars and 'fangs') makes it likely that their ancestor was a primitive creature with human dental traits. He suggests, drawing evidence from the Fayum fossil assemblage near Cairo, that the lines of man and ape separated nearly forty million years ago. It 'goes against the grain' that the man-like teeth of a cat-sized primate, *Propliopithecus*, would have evolved into the ape speciality, and then back into man-like teeth again.[5] So Kurten suggests, after the end of the Oligocene period, two separate lines of descent, one to apes, the other to man. He also says that there have been a number of different species of man. Such species may have coexisted but all except one, *Homo sapiens*, have now become extinct. In the section 'Bones of Contention', after we have asked whether an ape can create a sentence, we shall return to consider Kurten's idea.

Man possesses syntactical language, often more complex in construction in 'backward' societies than in our own. The study of the earliest known language, Sanskrit, revolutionized the study of language and grammar, giving rise to the science of comparative philology. Beyond the spoken, the written word has allowed man to develop the information banks which underwrite his technical and cultural achievements.

Of course, animals communicate; birds sing, bees dance, gorillas gesture and dolphins give whistles, clicks, creaks, squawks and other sound signals. And of course they can think. Cats, rats, chimps, pigeons etc. are able to recognize and respond to, if not always solve, contrived problems in mazes. What is unique to humans?

Animal signals are of limited dimensions and message, basically emotional and concerned with distress, food, sex etc. No clearly defined linear string of words, with grammatical structure, has ever been heard from an animal. Furthermore no animal can express, and then re-express in another way, an abstract idea. There is no great difference between the sounds of geographically separated members of the animal community. Wolves howl no matter where you hear them. With man language, among equally intelligent groups, varies greatly, so greatly that each group's language may be at first incomprehensible to the other. Moreover, words are symbols: they are a kind of code. Animals cannot manipulate symbols in the way linguistic or mathematical ability requires. In this way, a great chasm exists between the mind of man and ape.

It is true that, given a human teacher, animals may come to understand words and sign language as well as gestures and tones of voice. Apes are intelligent, sensitive creatures which respond to human affection. But this does not mean that, any more than parrots, they would by themselves have developed complex symbol-systems. It does not make them potential humans, any more than their humanoid form need have inexorably evolved into man. Only theory requires that.

The imitativeness of apes is remarkable. The husband-and-wife team of Dr R.A. Gardner and Dr B.T. Gardner began their famous experiments with Washoe, a chimpanzee, in June 1966, in association with the University of Nevada. They knew earlier experiments had shown that apes have relatively little interest in sound-play. Indeed, a previous experiment with Viki, a chimpanzee, had highlighted the quite serious psychological barriers against apes vocalizing. After six years she could use only four words and a few other sounds. There was no comprehension if previously learnt phrases were rearranged. Also, a chimp is unable to make refined use of its vocal chords. So the Gardners chose American sign language, as used by many deaf people in America.

After intensive care, Washoe mastered sixty-seven signs and 294 two-word permutations of these signs. Visually guided imitation brought rewards; but human language depends largely on auditory rather than visual factors. The human child has the capacity to imitate sounds but the chimpanzee's capacity is greatly inferior. Its auditory-based recall, essential to the acquisition of language, is less than a parrot's or crow's and neither creature has the cognitive processes of a human child. Apes merely learn, from the start, parrot-fashion. Children follow a pattern of language acquisition and grammatical regularization. By the age of two they will have passed the chimp and go on to learn an alphabet, spelling and writing. They may learn different languages and scripts but the pattern is the same. Washoe was a chasm away from human children as regards language potential.

Can an ape create a sentence? Studies have yielded no evidence of an ape's ability to use grammar. Such instances of presumed grammatical competence have been explained adequately by simple non-linguistic processes. The mean length of a child's sentence increases, so does its complexity. Not so with apes. Apes can, with non-simian tuition, pick up vocabularies of visual signals but there is no evidence that they can combine such symbols in order to create new meanings. They cannot construct syntactic sentences with different permutations of the same words. In contrast, by the age of five children with no special training can construct syntactical language. They can answer simple questions dealing with space and time, the measurement of heat, differences in size and weight, comparatives and superlatives, regular and irregular verb forms, pluralizations, active and passive moods, past and future tenses and much more. Elementary mathematical, imaginative and conceptual abilities are emerging.

It is psychologist Noam Chomsky's belief that all human languages share deep-seated properties of organization and structure in the brain.[6] These 'linguistic universals' are, he assumes, an innate mental endowment rather than the result of learning. If the capacity for language acquisition is innate in humans, may we presume the other forms of human quest (science and technology, philosophy and religion, literature etc.) are also exclusively human?

Dogs have a keener sense of smell, elephants keener hearing, eagles sharper sight than man. Organisms are equipped for different roles, express different characters in an ecological drama. In the play of life on earth, mankind, granted a keener intellect than all the rest, has thereby inevitably been cast in the lead role. For, above all senses and responses, it is mind which rules. The basis of human endeavour and achievement is language, both verbal and mathematical; in short, the ability to manipulate symbols. The creationist view, like Chomsky's, is that language is programmed into human being. There is absolutely no evidence that it evolved.

Bones of Contention

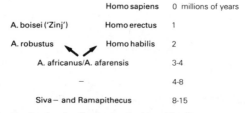

	Homo sapiens	0 millions of years
A. boisei ('Zinj')	Homo erectus	1
A. robustus	Homo habilis	2
A. africanus/A. afarensis		3-4
—		4-8
Siva— and Ramapithecus		8-15

Figure 39. *Hominid lineage showing the distribution in time of fossils – a recent view.* Note that up to four million years ago fossils are scarce; that there is a significant lack of hominid fossils between four and eight million years old; and that divergence is thought to have occurred from *A. afarensis* ('Lucy') to 'Zinj' – a line now extinct – and to yourself, *Homo sapiens.*

Kurten believes that about 35 million years ago stock diverged: *Propliopithecus* led to the hominids (as shown above) and *Aegyptopithecus* to pongids, including *Dryopithecus.* Others, such as Pilbeam, believe Miocene *Proconsul africanus* was the 'formative ape'. Then 16 million years ago, African and Asian apes split. Man derived not from Asian Siva- or Ramapithecines but from the African stock. After 10 million years ago first gorillas, then chimps and, by 4 million years, hominids split from ancestors

Bjorn Kurten believed that the 'common ancestor' of both apes and men was a creature like *Propliopithecus* of the Fayum although, recently, parallels have been drawn between it and the living primate, *Alouatta*, the howling monkey.[7] Following *Propliopithecus*, Kurten believed *Ramapithecus* was the late Miocene ancestor of *Homo sapiens.* This creature, a forest-dweller which became extinct about ten million years ago, was only four feet high and known, until 1977, by only about forty teeth and jaw fragments. Was it really an early hominid? Recent expeditions to the sub-Himalayan Siwalik hills in Pakistan have uncovered more 'pre-human' fossils of the very similar ramapithecine and sivapithecine varieties. A new skull, by far the most complete *Sivapithecus* yet found, has reinforced the minority view that these creatures were more like orang-utans than anything else.[8] On the other hand, a baboon, *Theropithecus gelada*, which inhabits the mountains of Ethiopia, has been compared with *Ramapithecus.* It has incisors and canines which are small relative to those of extant

African apes; also close-packed heavily worn cheek teeth, powerful masticatory muscles, a short deep face and other 'manlike' features possessed by *Ramapithecus*. But no one doubts it is a baboon.

Claims have been made for human finds in the Pliocene era. These include skulls found at Calaveras in California and Castenedolo in Italy. In the first case a gold-miner, digging in 1866, unearthed a human skull thickly encased in cemented gravel from 130 feet below the surface. But, along with others that were found, it was dismissed as an intrusive burial.

In 1860 at Castenedolo, Italy, Professor Razzagoni uncovered a human skull in Pliocene strata. Careful inspection of the overlying rocks showed them undisturbed. Sir Arthur Keith admitted: 'As the student of prehistoric man reads and studies the records of the Castenedolo finds, a feeling of incredulity rises within him. He cannot reject this discovery as false without doing injury to his sense of truth, and he cannot accept it as fact without shattering his accepted beliefs.'[9]

'Injury' was preferred to 'shattering' and Razzagoni's find was dismissed, but it is hard to dismiss the Vertesszöllös skull fragment from Hungary (*c*. 300,000 years ago) or the pair of specimens from the Fontéchevard cave in France. These are modern in aspect, 100,000 years old and found in 1947 by experienced excavators led by Germaine Henri-Martin. The age of the deposit was defined by the animal types it contained; because it had been sealed by a thick, undisturbed layer of stalagmites, later burial was impossible. What are we to believe?

Chimp or Pygmy?

What about the southern ape, *Australopithecus*? This is not so much a grey area as one hidden from objective view by the blinding light of revelation after revelation. They are regarded variously as apes, ape-men or members of an extinct sideline which, although emergent from a common ancestor, were thereafter never on the human line. In contrast, Kurten believed these metre-high creatures were a primitive species of man descended from *Ramapithecus* on his 'non-ape' line.

The confusion over *Australopithecus* is illustrated in the case of Professor Ashley Montagu, a leading American anthropologist. In 1957 he wrote that such an extremely ape-like creature could have nothing to do with man. In 1964, with Dr G. Loring Brace, he changed his mind and placed *Australopithecus* in the genus *Homo*. But by 1977 he had reclassified them again as the separate genus *Australopithecus*. Such vacillation does not cast doubt on the professional integrity of Dr Montagu; it highlights the insubstantial nature of the evidence.

In 1924 Raymond Dart discovered a juvenile skull at The Place of the Lion (Ta-ung) in southern Africa. This caused controversy because, although Dart pointed out many ape-like features, he believed that other features, particularly the teeth, were man-like. When he claimed it was hominid most contemporary workers disagreed, considering 'Dart's baby' a form of chimpanzee. Other bones were collected from Sterkfontein, Swartkrans and Makapansgat (also in South Africa). Professor Solly Zuckerman and experts on his staff investigated these bones for some years and came to the conclusion they were apes.

The crucial issue is 'Did *Australopithecus* walk upright?' In the public mind *Australopithecus* is closely linked with the Leakey family (working at Olduvai and other sites in Kenya and Tanzania) and Don Johanson in Ethiopia. They and others believe the 'apes' did walk upright, and can show footprints to prove it.

After Dr Louis Leakey's death in 1972, his wife Mary continued excavating both at Olduvai and further south at Laetoli. A few jaws were found and classified *Homo erectus* like Java and Peking man. She also discovered, amid many bird and animal tracks identical to modern types, a trail of ape-like footprints in volcanic ash considered to date from nearly four million years ago. The three Laetoli tracks were publicized in 1978. The big toes are not simian, i.e., opposable. From slides and photographs these tracks appear pretty much as yours or mine would if we walked barefoot across such a surface. They are 'free-striding' and indicate an upright gait with one foot crossing in front of the other in the manner of a hunter. They could easily have belonged to a modern girl or a pygmy. The shortish stride could have arisen because the person was walking cautiously across damp volcanic ash. Nevertheless, rather than 'human' the term 'hominid' is generally used in commenting on them, and the prints are interpreted as the tracks of a small-brained, upright ape.

Dr Johanson and his colleague Tim White believe that the Laetoli hominids are the same group (*Australopithecus afarensis*) as his finds at Hadar in the Afar region of Ethiopia. In 1972 Frenchman Maurice Taieb took Johanson to the region which Teilhard had visited in 1928 when returning to China from France. Expeditions since 1972 have unearthed various fragments of 'a family of hominids' said to have died in a flash-flood (but found in thirty-three feet of mudstone!). Among them is 'Lucy', a skeleton to reckon with. Though far from complete, Lucy is our most convincing australopithecine, found in 1974 and dated by Johanson as between 2.8 and 3.3. million years old. Johanson at first classed her as human in the genus *Homo* but by 1979 he had concluded that she was at the very base of human stock, pushing at the limits of hominid divergence from the apes. By everyone's admission the creature was very 'primitive' and its closest living representative would probably be the chimpanzee. The unbeliever, in spite of Johanson's conviction that this 'very primitive' ape-like skull was perched on a bipedal body, would note the missing hands and feet and crushed knee-joint. He would suggest it was not human at all but a species representing a form much closer to chimpanzee or gorilla. A complete skull, made up from fragments of different individuals, looked like a small female gorilla. And the palate is considerably more like that of a chimpanzee than a human.

In 1973 two bones of a knee-joint were found in a stratum, eighty metres lower than that in which Lucy was found in 1974. The actual distance between them is not given, but it is assumed they were from the same individual. The lower bone (tibia) shows, like Lucy's, no human characteristics. Lucy's upper bone (femur) is badly crushed at the knee-joint, the very part which ought to show bipedalism if it existed. The other upper bone was held to exhibit unquantified evidence of upright posture. Thus from the doubtful witness of one small bone found eighty metres deeper and some distance away, Lucy is assumed to have walked upright. But did she?

The evidence from arm-leg ratios is inconclusive, not least due to the condition of her limbs. The pelvis — which would be broad in a human — is said to have been distorted and to have undergone laboratory 'correction' to reduce the distortion. How can anyone know that the bones of a unique specimen are distorted, and by how much? How will they recognize the 'right' position and will this become further evidence for Lucy's bipedalism? And thus of her humanity? Does it matter anyway? The rain-forest chimpanzees *Pan paniscus*, spend a good deal of time walking upright, and no one would claim they were human.

Australopithecus, at four feet, never grew much larger than large chimpanzees. Dental analysis has shown that they were, like *Ramapithecus*, vegetarian. But if a group of them decided to swing down from the trees and become meat-eating *Homo erectus* on the plain, upright gait would be the last thing they would want. Their first efforts would give them an uncomfortable short-stretch roll, and a slow one at that. Man walks about as fast as a chicken; he runs upright at 12 m.p.h. while the patas monkey can run two-and-a-half times as fast. Indeed, the new man would have been about the slowest mammal on the savannah; rolling like a boat in high seas, and still wearing that tiny chimpish head, he'd have had little chance in the survival stakes. *Australopithecus*, if he wasn't some kind of chimp, didn't make it. After living side by side with modern man he is presumed to have become extinct about a million years ago.

Lucy became a star. But who was *Homo habilis*? Handy man (perhaps four feet high and 1.8 million years old) is still the subject of debate. Presumably on the basis of part of a toeless foot, Dr Leakey believed this creature walked upright. Several skull fragments (including one trampled into hundreds of pieces by cattle) and teeth were found in the lowest two beds at Olduvai gorge with what was interpreted as a bone tool and, a mile away, a rough circle of loosely piled stones construed as a windbreak, a temporary shelter such as hunters make. But Leakey's designation 'Homo' (man) was challenged. Even Leakey himself did not believe that *Homo habilis* could have been a transitional form between *Australopithecus* and alleged *Homo erectus*. Most anthropologists believed it was an australopithecine, whose profile can be superimposed (with a close fit) on that of a chimpanzee.

The manufactured tools need not have been made or even used by Leakey's *Homo habilis*, fragments, most of which were found only a few hundred yards from the site of another ex-star, *Zinjanthropus boisei* or, in common parlance, East African nutcracker man. 'Zinj', found by Mary Leakey at Olduvai in 1959, was the subject of great publicity by his sponsor, the National Geographic Society. Artists commissioned in the early sixties reconstructed 'identikit' portraits from the single skull – itself a reconstruction. Different hair styles were employed so that the 'man' ranged from apish to lowbrow human.

The skull was badly shattered and was accompanied by the fossilized bones of many animals which had been broken open for their marrow. This suggests that 'Zinj' might have served as supper for a more advanced type. It could not be a 'missing link' because it lived at the same time (about 1.75 million years ago) as other 'early fossil men'. It was found slightly above several habiline fragments. Anyway, the bony crest of Zinj's calvarium undoubtedly bespeaks apehood and by 1965 even Dr Leakey had conceded that it was most like a gorilla.

Before they died out, the australopithecines never changed in three million years. Their brains never evolved in size. Johanson writes that, by Richard Leakey's definition, modern chimpanzees would be classified as *A. africanus*.[10] Another kind, *A. robustus*, has more massive teeth and jaws and possesses the bony ridges found in gorillas and orang-utans. Was it, as has been suggested, the mate of *A. africanus*? For the teleologist, australopithecines are the stuff that dreams and fortunes are made on; in fact, different species, perhaps genera, of various apes.

The Fount of Wisdom

Darwin wrote the *Descent of Man* and Bronowski *The Ascent of Man*. Up or down, right or wrong, both were evolutionists. But, whether created or evolved, man presumably diverged from one centre of origin. Where? Whence did he descend? It is hard to see the wood for the trees!

Ramapithecus, scant remains of which are claimed from Turkey to China, is out. Java and Peking men are suspect. The centre of gravity has swung towards Africa and, specifically, Lucy and the Laetoli 'hominids'.

Despite pronounced differences in shape, the great apes and men differ only in about 1 per cent of their DNA. Chimp and man are genetically closer than zebra and horse, which are estimated to have diverged only five million years ago. In this case, the man-chimp divergence should be even younger. This view sees Lucy at the point of hominid divergence from the apes; bipedalism and not brain size is the critical discontinuity which triggered the human 'shot in the dark'. This theory fits with Lucy's tiny head.

Biochemical and immunological data suggests, in contrast to Kurten's fossil-based speculation, a common ancestor for the gorilla-chimp-man cluster at about five million years. It has also been noted that no fossils have been found unequivocally ancestral to chimp and gorilla but not to man. Non-palaeontologists have speculated that *Australopithecus*, extinct for around a million years, is now represented by the chimp and gorilla for which there are no fossil ancestors.[11] Brain sizes are similar. Lucy's composite skull resembled a small gorilla. The chimp is therefore supposed to represent a *reversion* from a more man-like ancestor.

That such ideas can even be mooted indicates the state of play. It is a mere whisper from the creationist position that *Australopithecus* was never other than distinctly ape.

Is this how far we have come since Darwin, over a hundred years ago, regretted 'missing links'? A campfire, flinty clicks as tools are chipped, the meal or, as they trek to hunt, a silhouette of ape-like pygmies in the human dawn. It is easy to clothe insubstantial ape-man in a thin web of story-telling. Was naked skin favoured by natural selection because its intimacy was greater than that of furred skin — while genital and armpit hair remained as sexual signals and scent traps? Did face-to-face matings increase among *Australopithecus*? Was it as a prelude to marriage that lips became full, breasts enlarged and, as a harbinger of morality or monogamy, or both, virgin ape-girls evolved the hymen? Sociobiologies of prehistoric family life and love have indeed been evolved from a few sticks and bones.[12]

If man was created, we do not know where. But if man was created with woman, an interesting fact has been pointed out by Francisco Ayala, a Professor of Genetics at the University of California.[13] He believes that about 7 per cent of human genes show alleles. In other words, this proportion of genes can show variations in the same character, for example, straight versus curly hair, brown versus blue or green eyes etc. On the basis of this 'heterozygosity' Professor Ayala has calculated that the average human couple could have 10^{2017} children before they had one child identical to another. This number is far greater than the stars in the sky, the grains of sand by the sea, all creatures born or atoms in the universe (about 10^{80}).

Take skin colour. Everyone except albinos has the same skin colouring, a protein called melanin. We just have different amounts of this basic colour; and the amount depends on at least two pairs of genes. We can call these P and Q. Remember that humans and most other organisms have two sets of similar chromosomes, one from their mother and one from their father. So a person will have one P gene from mum, one from dad. Similarly with Q.

	PQ	Pq	pQ	pq
PQ	PP QQ	PP Qq	Pp QQ	Pp Qq
Pq	PP Qq	PP qq	Pp Qq	Pp qq
pQ	Pp QQ	Pp Qq	pp QQ	pp Qq
pq	Pp Qq	Pp qq	pp Qq	pp qq

Figure 40. *Maximum variation for PpQq × PpQq*, that is, combinations produced when male and female both with the genotype PpQq for skin colour are mated

In fact there are at least four skin-colour genes in the human gene-pool: P, p, Q, q. While P and Q code for dark skin, p and q are alleles that code for light skin. It is therefore possible for a person's colour complement to be PPQQ (dark skin), PpQq (medium skin) or ppqq (light skin).

Starting with two medium-skinned parents, PpQq, it is possible to achieve a maximum number of permutations shown in the diagram.

In other words, in one generation this couple could produce the whole range of skin traits. There is only about one chance in sixteen of either negroid (PPQQ) or Scandinavian (ppqq) coloration. The rest would span the intermediate range.

After the first generation, if those with dark skins migrated into the same areas and/or took only dark mates, than an 'upper case' syndrome (PPQQ) would occur. Potential would have been reduced and offspring would always be black. Similarly, very fair parents (ppqq), having no 'upper case' genes to pass on, would be locked into the production of pale children.

If people with different skin colours reunited, however (as occurs, for example, in the West Indies), then both 'upper' and 'lower case' genes can mix and realize their original potential. The full range of variation can, and does, occur again.

This is an example of how only two simple pairs of genes concerning one trait (skin colour) can permutate. Because, like colours on a rubic cube, ordered potential was present from the start, it would not take long for diverse permutations to occur. By multiplying many such traits, Ayala has arrived at his staggering figures which indicate the genetic potential humans (and, clearly, other organisms) possess.

Some genes, of course, do not express themselves in 'ranges' of degree. It is all or nothing. A characteristic is either expressed or suppressed: but not lost. The creationist believes that archetypes incorporate, in the form of created alleles, the potential for innumerable permutations. Wear and tear, represented by non-lethal mutations, in time adds to the number of alleles — like dents or scratches on a motor car.

In the last analysis, humans, perhaps created in the watered parks of India, the Middle East or even Southern Africa, *could* have diversified and migrated out, as wandering tribes, to all the habitable corners of earth. Did the early races or species of man look and behave like stone-age natives do today — whether bushmen, Aborigines or Amazonian Indians? How well would you be coping after a few months if you were 'loosed', with all remembrance of technology and civilization erased from your mind, onto virgin continent? It would be a fresher start than you or I can imagine — an Adam or an Eve. If you and a mate survived you might generate peoples who, as they migrated from the original centre, developed their individual technologies and cultures. After many an epoch one of these civilizations might even come to surpass that of the society from which, millenia previously, you had been unwittingly 'loosed'.

The Upshot

The creationist thesis is clear. No son of ape rose into Tarzan. Man is a distinct type. He did not diverge from an evolutionary trunk: his tree grew straight up from its own roots. Apes and humans existed in the past as they do in the present, as separate types with large but limited variability. Fossil evidence substantiates the creationist's case, i.e., that fossils proposed as 'missing links' to illustrate the evolution of mankind are in reality (when they are not frauds) either men or apes. They do not sit easily somewhere between.

From pygmies to 'giants', from Chinamen to Scandinavians different permutations of human genetic potential have congregated. The creationist expects this 'micro-evolution' to have occurred, in the past as in the present, around the human archetype; his view could extend to Himalayan yetis, Canadian 'bigfoots', Caucasian almas or any other creature that shows human characters, however different the form in culture, however hairy, however 'Neanderthal' in appearance and dress. For the creationist, degeneration may have occurred in some human types, but never did Adam evolve; nor will Eve evolve into superwoman.

I started as devil's advocate for the creationist view and came, in principle, though not according to any particular creed, to prefer it. This book drives the thin end of a philosophical wedge into materialism. For this it will probably be condemned from some quarters. But I hope I have shown that apparently convincing arguments in support of a belief can often be seen to be either based on insufficient data or open to more than one interpretation; and that much of what passes for science is no more and no less emotional, illogical and idiosyncratic than many of the opposing arguments.

Science, useful as it is, does not explain a host of things; nor is all that it does not explain false. Your own feelings are an example. These are the primary data of psychology, the basis of your experience of life – yet they are not well understood in terms of science. The hard-working, scientific approach is worthy and progressive but is perhaps insufficient to explain the origin of life.

Nevertheless, many scientists have taken the simple concepts proposed by Darwin and waltzed off with them. It is as if they repeated a conceptual mantra. They become blind to imperfections and make a virtue of accepting the shortcomings of the theory of evolution (the geological record, for example) in the faith that their tolerance will be rewarded. A man's gospel is his business: that he teaches evolution as holy writ in television series or in schools and colleges – with no alternatives properly considered – is a more serious matter.

While it entirely accepts science and its methods, the creationist view has roots beyond the materialistic vision of science. I personally have had no inward revelations as to how or why creation occurred. Did a complete creation or successive creations occur? Was life on earth made once-for-all or was its complexity purposely increased by stages on an age-old earth? Perhaps we will never know. What matters is the fact of coded *design* and the sensible inference of *creation* that it is possible to draw from a study of the biological machines we call plants and animals.

Because in our age there is no doubt that doubt exists, I believe this inference of creation must be clearly and fairly expressed. Equally, in terms of the same scientific data, the inference of evolution needs to be expressed. Presenting one viewpoint exclusively is faulty teaching. Our descendants will marvel at the attempts of the neo-Darwinian lobby to suppress alternative inquiry, as we today marvel at the power of churchmen in the eighteenth and nineteenth centuries.

Adam and Evolution should be controversial. The many issues it raises cannot all be dealt with, let alone in depth, in a single sweep. But the direction of the argument is clear – there has been neither chemical evolution nor macro-evolution. Nor, as some twentieth century churchmen bio-illogically accept, did God involve chance mutations in 'creation by evolution'. No intelligent creator would leave matters to chance; on the contrary, his purpose would be to realize, in plan and in practice, his ideas. Pressing the logic to its conclusion, this book advocates a grand and full-blooded creation. The implications of this view necessitate a reappraisal of ourselves and of the whole world of organisms around us.

Notes

1 Lewin, R., *A New Focus for African Prehistory*, New Scientist, vol. 75, no. 107, 29 September 1977, p. 793.
2 Leakey, R., and Lewin, R., *Origins*, MacDonald and Jane's, 1977, p. 106.
3 Barclay, V., *Darwin is Not for Children*, Herbert Jenkins, 1950, esp. chapter 14.
4 Montagu, A., *Growing Young*, McGraw Hill, 1981, p. 16.
5 Kurten, B., *Not from the Apes*, Gollancz, 1972, pp. 38–40.
6 Chomsky, N., *Psychology Today*, February 1965, pp. 432–3.
7 Fleagle, J.G., 'Ape Limb Bone from the Oligocene of Egypt', *Science*, vol. 189, no. 4197, 11 July 1975, p. 136.
8 Andrews, P. 'Humanoid Evolution', *Nature*, vol. 295, 21 January 1982, p. 185.
9 Keith, A., *The Antiquity of Man*, Williams and Norgate, 1925, p. 334.
10 Johanson, D., *Science*, vol. 207, 7 March 1980, p. 1105.
11 Cherfas, J., and Gribbin, J., 'Descent of Man – or Ascent of Ape', *New Scientist*, vol. 91, no. 1269, 3 September 1981, pp. 592–5.
12 Kurten, B., op. cit., p. 94.
13 Ayala, F., 'The Mechanisms of Evolution', *Scientific American*, vol. 239, no. 3, September 1978, p. 55.

Glossary

The Uniformitarian Geological Timetable

Era	Period	Estimated start/millions of years
	Quaternary:	
	Recent epoch	0.1
	Pleistocene	2.0
	Tertiary:	
CENOZOIC	Pliocene	5
	Miocene	22.5
	Oligocene	36
	Eocene	55
	Palaeocene	65
MESOZOIC	Cretaceous	135
	Jurassic	190
	Triassic	225
PALAEOZOIC	Permian	280
	Carboniferous	345
	Devonian	395
	Silurian	435
	Ordovician	500
	Cambrian	570
PROTEROZOIC		2500
	pre-Cambrian	
ARCHAEOZOIC		4500

Allele: two genes situated at the same point on a pair of homologous chromosomes are said to be alleles if they produce different effects on the same set of developmental processes. For example, one allele might cause eyes to be coloured brown, another blue.

Archetype: a prototype, ideal or fundamental structure common to organisms at molecular, cellular, organic or phenotypic levels. Also, the basic pattern or body-plan of a major taxonomic group.

Chloroplast: organelle in plant cells containing photosynthetic apparatus.

Chromosome: numbers of linear, threadlike bodies made of DNA and protein which are found in the nucleus of every eukaryotic cell. A single circular structure occurs in prokaryotes.

Ciliate: one of a group of unicellular organisms, including *Paramecium*.

Codon: the unit of DNA, made up of three nucleotides, which codes for one amino acid.

DNA: the material of inheritance, whose two strands of nucleotides are linked parallel to each other by base-pairing and coiled into a helix. DNA is found only in cells, in the nuclei of eukaryotes and the cytoplasm of prokaryotes.

Enzyme: a protein which catalyzes biochemical reactions in a body. Body chemistry (metabolism) is entirely dependent on enzymes; in most cases suites of enzymes, one for each particular reaction, cooperate in a multistage pathway towards the required end-result.

Eukaryote: any organism except bacteria and blue-green algae.

Gamete: a sex cell, egg or sperm, containing a single set of chromosomes.

Gametophyte: the sexually reproductive phase in the life-cycle of a plant.

Gene: the basic unit of material inheritance, possessing coded information for the production of protein.

Genetics: the study of genetic material and heredity.

Genome: the whole set of chromosomes in an organism.

Genotype: the genetical constitution of an organism.

Heterozygote: two genes at the same point on a pair of homologous chromosomes are said to be heterozygous, i.e., alleles, if they are different.

Homozygote: two genes at the same point on a pair of homologous chromosomes are said to be homozygous if they are identical.

Ion: a charged atom or group of atoms, that is, with more or less electrons than protons. Many crystals, like table salt, dissolve into ions.

Metazoan: a many-celled, as opposed to a single-celled, organism.

Mitochondrion: organelle in eukaryotic cells containing the apparatus for aerobic respiration.

Morphogenesis: the development of shape and form.

Nucleotide: DNA is a long chain of nucleotides. Free nucleotides are found in ATP, some coenzymes etc. They are composed of a sugar, phosphoric acid and a nitrogen-containing base (adenine, cytosine, guanine, thymine or uracil).

Ontogeny: developmental history, the course of development in an individual's life-cycle.

Organelle: an organelle, operating within a cell, is analogous to an organ in a whole organism. Examples are the nucleus, mitochondria or chloroplasts.

Palaeontology: the study of fossils.

Phenotype: the outward appearance of an organism, as opposed to its genotype.

Photosynthesis: the method by which plants convert light into chemical energy in the form of starch.

Phylogeny: evolutionary history, relationships based on evolutionary descent.

Polysaccharide: a carbohydrate made of a chain or a branched chain of sugar molecules. For example, starch is made of glucose molecules.

Prokaryote: blue-green alga or bacterium.

Protein: a peptide is formed when two or more amino acids are linked. A polypeptide is a chain of such acids and a functional polypeptide, such as egg-white, pepsin or gelatin is called a protein.

Ribosome: in the cell cytoplasm, the 'tape-head' unit through which m-RNA 'tapes' from the nucleus are run in order to translate them into proteins.

RNA: m-RNA forms a 'negative' of the DNA template from which it is peeled and carried to the ribosome. Each t-RNA molecule, which has a 'clover-leaf' shape, collects a specific amino acid on one end and with the other locks tight onto a specific codon of m-RNA. It is the translator, without which the genetic message could not be 'developed' as protein. Ribosomes are partly composed of r-RNA. In all its forms RNA is concerned to translate the structure of DNA into the structure of protein molecules.

Respiration: the controlled release of energy from food.

Sporophyte: the asexually reproductive (spore-producing) phase in the life-cycle of a plant.

Stoma (pl. Stomata): a mouth-like pore, surrounded by lip-like 'guard cells', which allow gases to diffuse in and out of a leaf.

Unconformity: occurs where, over a period of time, the deposition of sediment has been interrupted or rocks eroded into a landscape, leaving older rocks exposed and later buried by newer rocks. In either case there is a 'time-gap' between upper and lower layers.

Zygote: a fertilized egg.

Topical Reading

Ambrose, E.J., *The Nature and Origins of the Biological World*, Ellis Horwood, 1982.

Gould, S.J., *Ever Since Darwin*, Norton, 1979.

Gould, S.J., *Hen's Teeth and Horse's Toes*, Norton, 1983.

Gould, S.J., *The Panda's Thumb*, Norton, 1982.

Grassé, P.-P., *The Evolution of Living Organisms*, Academic Press, 1978.

Kerkut, G.A., *The Implications of Evolution*, Pergamon Press, New York, 1960.

Morris, H., ed., Scientific Creationism, Creation-Life Publishers, 1974.

Patterson, C., *Evolution*, Routledge and Kegan Paul, 1978.

Shute, E., *Flaws in the Theory of Evolution*, Craig Press, 1962.

Thorpe, W.E., *Purpose in a World of Chance*, Oxford University Press, 1978.

Utt, R. (ed.), *Creation, Nature's Design and Designer*, Pacific Press Publishing Association, 1971.

Chapter 2

Kitcher, P., *Abusing Science – the Case against Creationism*, MIT Press, 1982.

Chapter 3

Marsh, F., *Variation and Fixity in Nature*, Pacific Press Publishing Association, 1976.

Chapter 4

Watson, J., *The Double Helix*, Penguin, July 1970.

Chapter 5

Macbeth, N., *Darwin Retried – an Appeal to Reason*, Gambit, 1979.

Chapter 6

Bowden, M., *Ape-Man, Fact or Fallacy?*, Sovereign Publications, 1981.

Chapter 7

Bristow, A., *The Sex Life of Plants*, Barrie and Jenkins, 1978.

Chapter 8

Burr, H.S., *Blueprint for Immortality*, Spearman, 1972.

Koestler, A., *The Ghost in the Machine*, Picador, 1975. .

Sheldrake, R., *A New Science of Life*, Paladin, 1983.

Waddington, C.H., *The Nature of Life*, Harper Torchbook, New York, 1966.

Chapters 9, 10, 11

Aw, S., *Chemical Evolution*, University Education Press, Singapore, 1976.

Crick, F., *Life Itself*, MacDonald, 1982.

Hoyle, F., and Wickramasinghe, N., *Evolution from Space*, Dent, 1981.

Wilder Smith, A.E., *The Creation of Life – a Cybernetic Approach to Evolution*, Harold Shaw Publishers, Illinois, 1970.

Chapters 12, 13, 14, 15

Dewar, D., *The Transformist Illusion*, Dehoff Publications, Tennessee, 1957.

Stanley, S. M., *Macro-evolution – Pattern and Process*, W. H. Freeman and Company, San Francisco, 1979.

Chapter 16

Clark, R.E.D., *The Universe, Plan or Accident?*, Paternoster Press, 1961.

Hayward, A., *Creation – The Facts and the Fallacies*, SPCK, 1984.

Hoyle, F., *The Intelligent Universe*, Michael Joseph, 1983.

Lovell, B., *In the Centre of Immensities*, Paladin, 1980.

Morris, H., and Parker, G., *What is Creation Science?*, Creation-Life Publishers, California, 1982.

Chapter 17

Bowden, M., op. cit.

Kurten, B., *Not from the Apes*, Victor Gollancz, London, 1972.

Index

Acknowledgements

I should like to give special thanks to Drs Robert E.D. Clark and Bernard Stonehouse, whose detailed comments have much improved the presentation of the book.

The author and publishers wish to thank: The Academic Press for permission to quote from *The Evolution of Living Organisms* by Pierre-P. Grassé; Creation Life Publishers for permission to print the table shown in fig. 38; J.M. Dent for permission to quote from *Evolution from Space* by Sir Fred Hoyle and C. Wickramasinghe; Edward Arnold (Publishers) Ltd for permission to adapt the format of fig. 1.2 from M. Tribe and P. Whittaker's *Chloroplasts and Mitochondria* in the Studies in Biology series no. 31 — used in fig. 28; Garland Publishing Inc. for permission to adapt figs. 8.119, 9.19 and 14. 9 from Watson, *The Molecular Biology of the Cell* — used in figs. 27, 25 and 8; Thomas Nelson and Sons Ltd for the quote from *Biology — A Functional Approach* by Dr M.B.V. Roberts; New Scientist for quotes from Drs R. Lewin and M. Land; Octopus Books for permission to use the figures from *The Encyclopaedia of the Animal World* in figs. 31 and 33.